Student Solutions Manual

for

Devore's

Probability and Statistics
For Engineering and the Sciences

Seventh Edition

Matthew A. Carlton
California Polytechnic State University
San Luis Obispo

BROOKS/COLE
CENGAGE Learning

Australia • Brazil • Japan • Korea • Mexico • Singapore • Spain • United Kingdom • United States

BROOKS/COLE
CENGAGE Learning™

For product information and technology assistance, contact us at
Cengage Learning Customer & Sales Support, 1-800-354-9706

For permission to use material from this text or product, submit all requests online at
www.cengage.com/permissions
Further permissions questions can be emailed to
permissionrequest@cengage.com

ISBN-13: 978-0-495-38219-5
ISBN-10: 0-495-38219-1

Cover image: © Creatas/SuperStock

Brooks/Cole
10 Davis Drive
Belmont, CA 94002-3098
USA

Cengage Learning is a leading provider of customized learning solutions with office locations around the globe, including Singapore, the United Kingdom, Australia, Mexico, Brazil, and Japan. Locate your local office at:
www.cengage.com/global

Cengage Learning products are represented in Canada by Nelson Education, Ltd.

To learn more about Brooks/Cole, visit
www.cengage.com/brookscole

Purchase any of our products at your local college store or at our preferred online store
www.ichapters.com

Printed in the United States of America
2 3 4 5 6 7 11 10 09 08

CONTENTS

CHAPTER 1

Section 1.1

1.
    a. Houston Chronicle, Des Moines Register, Chicago Tribune, Washington Post

    b. Capital One, Campbell Soup, Merrill Lynch, Pulitzer

    c. Bill Jasper, Kay Reinke, Helen Ford, David Menedez

    d. 1.78, 2.44, 3.5, 3.04

3.
    a. In a sample of 100 computers, what are the chances that more than 20 need service while under warranty? What are the chances than none need service while still under warranty?

    b. What proportion of all computers of this brand and model will need service within the warranty period?

5.
    a. No, the relevant conceptual population is all scores of all students who participate in the SI in conjunction with this particular statistics course.

    b. The advantage to randomly allocating students to the two groups is that the two groups should then be fairly comparable before the study. If the two groups perform differently in the class, we might attribute this to the treatments (SI and control). If it were left to students to choose, stronger or more dedicated students might gravitate toward SI, confounding the results.

    c. If all students were put in the treatment group there would be no results with which to compare the treatments.

7.   One could generate a simple random sample of all single family homes in the city or a stratified random sample by taking a simple random sample from each of the 10 district neighborhoods. From each of the homes in the sample the necessary variables would be collected. This would be an enumerative study because there exists a finite, identifiable population of objects from which to sample.

1

9.

    a.   There could be several explanations for the variability of the measurements. Among them could be measuring error, (due to mechanical or technical changes across measurements), recording error, differences in weather conditions at time of measurements, etc.

    b.   This could be called an analytic study because there is no sampling frame.

Section 1.2

11.

```
6l |034
6h |667899
7l |00122244
7h |                        Stem=Tens
8l |001111122344            Leaf=Ones
8h |5557899
9l |03
9h |58
```

This display brings out the gap in the data: There are no scores in the high 70's.

13.

   a.

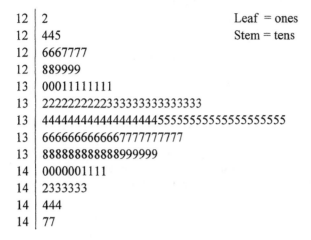

12	2
12	445
12	6667777
12	889999
13	00011111111
13	2222222222333333333333333
13	444444444444444444455555555555555555555
13	66666666666677777777777
13	888888888888999999
14	0000001111
14	2333333
14	444
14	77

Leaf = ones
Stem = tens

The observations are highly concentrated at $134 - 135$, where the display suggests the typical value falls.

   b.

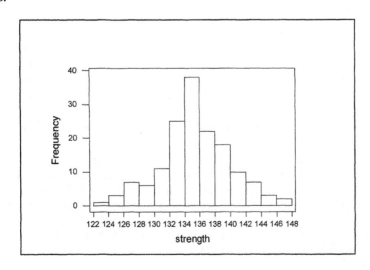

The histogram is symmetric and unimodal, with the point of symmetry at approximately 135.

15.

Crunchy		Creamy
	2	2
644	3	69
77220	4	145
6320	5	3666
222	6	258
55	7	
0	8	

Both sets of scores are reasonably spread out. There appear to be no outliers. The three highest scores are for the crunchy peanut butter, the three lowest for the creamy peanut butter.

17.

a.

Number Nonconforming	Frequency	RelativeFrequency(Freq/60)
0	7	0.117
1	12	0.200
2	13	0.217
3	14	0.233
4	6	0.100
5	3	0.050
6	3	0.050
7	1	0.017
8	1	0.017

doesn't add exactly to 1 because relative frequencies have been rounded 1.001

b. The number of batches with at most 5 nonconforming items is 7+12+13+14+6+3 = 55, which is a proportion of 55/60 = .917. The proportion of batches with (strictly) fewer than 5 nonconforming items is 52/60 = .867. Notice that these proportions could also have been computed by using the relative frequencies: e.g., proportion of batches with 5 or fewer nonconforming items = 1– (.05+.017+.017) = .916; proportion of batches with fewer than 5 nonconforming items = 1 – (.05+.05+.017+.017) = .866.

c. The following is a Minitab histogram of this data. The center of the histogram is somewhere around 2 or 3 and it shows that there is some positive skewness in the data. Using the rule of thumb in Exercise 1, the histogram also shows that there is a lot of spread/variation in this data.

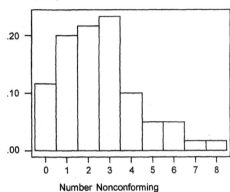

Number Nonconforming

19.

a. From this frequency distribution, the proportion of wafers that contained at least one particle is $(100-1)/100 = .99$, or 99%. Note that it is much easier to subtract 1 (which is the number of wafers that contain 0 particles) from 100 than it would be to add all the frequencies for 1, 2, 3,… particles. In a similar fashion, the proportion containing at least 5 particles is $(100 - 1-2-3-12-11)/100 = 71/100 = .71$, or, 71%.

b. The proportion containing between 5 and 10 particles is $(15+18+10+12+4+5)/100 = 64/100 = .64$, or 64%. The proportion that contain strictly between 5 and 10 (meaning strictly *more* than 5 and strictly *less* than 10) is $(18+10+12+4)/100 = 44/100 = .44$, or 44%.

5

c. The following histogram was constructed using Minitab. The data was entered using the same technique mentioned in the answer to exercise 8(a). The histogram is *almost* symmetric and unimodal; however, it has a few relative maxima (i.e., modes) and has a very slight positive skew.

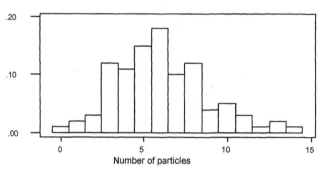

21.

a. A histogram of the y data appears below. From this histogram, the number of subdivisions having no cul–de–sacs (i.e., y = 0) is 17/47 = .362, or 36.2%. The proportion having at least one cul–de–sac (y ≥ 1) is (47–17)/47 = 30/47 = .638, or 63.8%. Note that subtracting the number of cul–de–sacs with y = 0 from the total, 47, is an easy way to find the number of subdivisions with y ≥ 1.

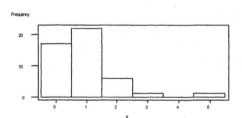

b. A histogram of the z data appears below. From this histogram, the number of subdivisions with at most 5 intersections (i.e., $z \leq 5$) is 42/47 = .894, or 89.4%. The proportion having fewer than 5 intersections ($z < 5$) is 39/47 = .830, or 83.0%.

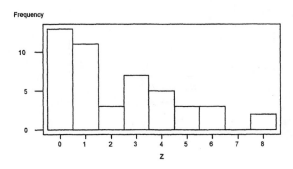

23.

a.

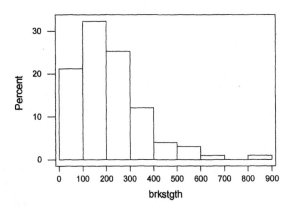

The histogram is skewed right, with most observations between 0 and 300 cycles. The class holding the most observations is between 100 and 200 cycles.

b.

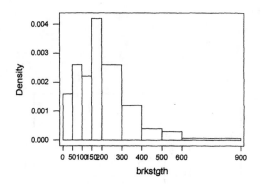

c   [proportion ≥ 100] = 1 – [proportion < 100] = 1 – .21 = .79

25.   Histogram of original data:

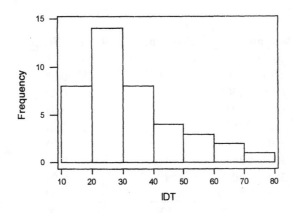

Histogram of transformed data:

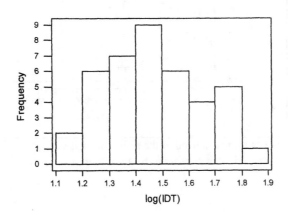

The transformation creates a much more symmetric, mound–shaped histogram.

27.

a. The endpoints of the class intervals overlap. For example, the value 50 falls in both of the intervals 0 – 50 and 50 – 100.

b.

Class Interval	Frequency	Relative Frequency
0 – < 50	9	0.18
50 – < 100	19	0.38
100 – < 150	11	0.22
150 – < 200	4	0.08
200 – < 250	2	0.04
250 – < 300	2	0.04
300 – < 350	1	0.02
350 – < 400	1	0.02
>= 400	1	0.02
	50	1.00

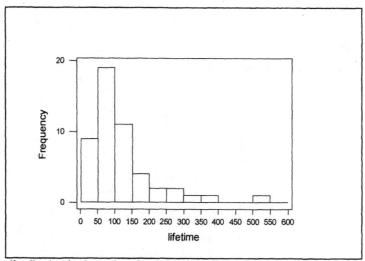

The distribution is skewed to the right, or positively skewed. There is a gap in the histogram, and what appears to be an outlier in the 500 – 550 interval.

c.

Class Interval	Frequency	Relative Frequency
2.25 – < 2.75	2	0.04
2.75 – < 3.25	2	0.04
3.25 – < 3.75	3	0.06
3.75 – < 4.25	8	0.16
4.25 – < 4.75	18	0.36
4.75 – < 5.25	10	0.20
5.25 – < 5.75	4	0.08
5.75 – < 6.25	3	0.06

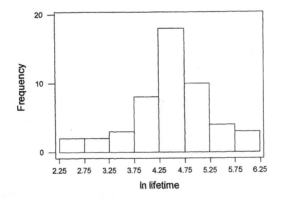

The distribution of the natural logs of the original data is much more symmetric than the original.

d. The proportion of lifetime observations in this sample that are less than 100 is .18 + .38 = .56, and the proportion that is at least 200 is .04 + .04 + .02 + .02 + .02 = .14.

29.

Complaint	Frequency	Relative Frequency
B	7	0.1167
C	3	0.0500
F	9	0.1500
J	10	0.1667
M	4	0.0667
N	6	0.1000
O	21	0.3500
	60	1.0000

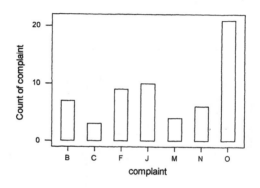

31.

Class	Frequency	Cumulative Frequency	Cumulative Relative Frequency
0.0 – under 4.0	2	2	0.050
4.0 – under 8.0	14	16	0.400
8.0 – under 12.0	11	27	0.675
12.0 – under 16.0	8	35	0.875
16.0 – under 20.0	4	39	0.975
20.0 – under 24.0	0	39	0.975
24.0 – under 28.0	1	40	1.000

Section 1.3

33.

    a. $\bar{x} = 192.57$, $\tilde{x} = 189$. The mean is larger than the median, but they are still fairly close together.

    b. Changing the one value, $\bar{x} = 189.71$, $\tilde{x} = 189$. The mean is lowered, the median stays the same.

    c. $\bar{x}_{tr} = 191.0$. $1/14 \approx .07$, or 7% trimmed from each tail.

    d. For $n = 13$, $\Sigma x = (119.7692) \times 13 = 1,557$. For $n = 14$, $\Sigma x = 1,557 + 159 = 1,716$. $\bar{x} = 1716/14 = 122.6$.

35.

    a. The sample mean is $\bar{x} = (100.4/8) = 12.55$.

        The sample size ($n = 8$) is even. Therefore, the sample median is the average of the ($n/2$) and ($n/2$) + 1 values. By sorting the 8 values in order, from smallest to largest: 8.0 8.9 11.0 12.0 13.0 14.5 15.0 18.0, the forth and fifth values are 12 and 13. The sample median is $(12.0 + 13.0)/2 = 12.5$.

        The 12.5% trimmed mean requires that we first trim $(.125)n$ or 1 value from the ends of the ordered data set. Then we average the remaining 6 values. The 12.5% trimmed mean $\bar{x}_{tr(12.5)}$ is $74.4/6 = 12.4$.

        All three measures of center are similar, indicating little skewness to the data set.

    b. The smallest value (8.0) could be increased to any number below 12.0 (a change of less than 4.0) without affecting the value of the sample median.

    c. The values obtained in part (a) can be used directly. For example, the sample mean of 12.55 psi could be re–expressed as $(12.55 \text{ psi}) \times \left(\dfrac{1ksi}{2.2\,psi} \right) = 5.70ksi$

37. $\bar{x} = 12.01$, $\tilde{x} = 11.35$, $\bar{x}_{tr(10)} = 11.46$. The median or the trimmed mean would be good choices because of the outlier 21.9.

39.

    a. $\Sigma x = 16.475$, so $\bar{x} = \dfrac{16.475}{16} = 1.0297$. $\tilde{x} = \dfrac{(1.007 + 1.011)}{2} = 1.009$.

    b.   1.394 can be decreased until it reaches 1.011(the largest of the 2 middle values) – i.e. by $1.394 - 1.011 = .383$. If it is decreased by more than .383, the median will change.

41.

    a. $7/10 = .70$

    b. $\bar{x} = .70$ = proportion of successes

    c. $s/25 = .80$, so $s = (0.80)(25) = 20$ successes total, so $20 - 7 = 13$ of the new cars would have to be successes.

43.   median $= \dfrac{(57 + 79)}{2} = 68.0$, 20% trimmed mean = 66.2, 30% trimmed mean = 67.5.

Section 1.4

45.

a. $\bar{x} = 577.9/5 = 115.58$. Deviations from the mean:
$116.4 - 115.58 = .82, 115.9 - 115.58 = .32, 114.6 - 115.58 = -.98,$
$115.2 - 115.58 = -.38,$ and $115.8 - 115.58 = .22.$

b. $s^2 = [(.82)^2 + (.32)^2 + (-.98)^2 + (-.38)^2 + (.22)^2]/(5-1) = 1.928/4 = .482$, so $s = .694$.

c. $\sum_i x_i^2 = 66,795.61$, so $s^2 = \frac{1}{n-1}\left[\sum_i x_i^2 - \frac{1}{n}\left(\sum_i x_i\right)^2\right] =$
$[66,795.61 - (577.9)^2/5]/4 = 1.928/4 = .482.$

47. The sample median is $\tilde{x} = 114$ the sample mean is $\bar{x} = 1,162/10 = 116.2$.
The sample standard deviation is

$$s = \sqrt{\frac{\sum x_i^2 - \frac{\left(\sum x_i\right)^2}{n}}{n-1}} = \sqrt{\frac{140,992 - \frac{(1,162)^2}{10}}{9}} = 25.75.$$

On average, we would expect a fracture strength of 116.2. In general, the size of a typical deviation from the sample mean (116.2) is about 25.75. Some observations may deviate from 116.2 by more than this and some by less.

49.

a. $\Sigma x = 2.75 + ... + 3.01 = 56.80$, $\Sigma x^2 = (2.75)^2 + ... + (3.01)^2 = 197.8040$

b. $s^2 = \dfrac{197.8040 - (56.80)^2/17}{16} = \dfrac{8.0252}{16} = .5016$, $s = .708$

51.

a. $\Sigma x = 2563$ and $\Sigma x^2 = 368,501$, so $s^2 = \dfrac{[368,501 - (2563)^2/19]}{18} = 1264.766$
and $s = 35.564$

b. If y = time in minutes, then $y = cx$ where $c = \frac{1}{60}$, so

$$s_y^2 = c^2 s_x^2 = \frac{1264.766}{3600} = .351 \text{ and } s_y = cs_x = \frac{35.564}{60} = .593$$

53.

lower half: 2.34 2.43 2.62 2.74 2.74 2.75 2.78 3.01 3.46
upper half: 3.46 3.56 3.65 3.85 3.88 3.93 4.21 4.33 4.52
Thus the lower fourth is 2.74 and the upper fourth is 3.88.

a. $f_s = 3.88 - 2.74 = 1.14$

b. f_s wouldn't change, since increasing the two largest values does not affect the upper fourth.

c. By at most .40 (that is, to anything not exceeding 2.74), since then it will not change the lower fourth.

d. Since n is now even, the lower half consists of the smallest 9 observations and the upper half consists of the largest 9. With the lower fourth = 2.74 and the upper fourth = 3.93, $f_s = 1.19$.

55.

a. Lower half of the data set: 325 325 334 339 356 356 359 359 363 364 364 366 369, whose median, and therefore the lower quartile, is 359 (the 7th observation in the sorted list).
The top half of the data is 370 373 373 374 375 389 392 393 394 397 402 403 424, whose median, and therefore the upper quartile is 392. So, the IQR = 392 − 359 = 33.

b. 1.5(IQR) = 1.5(33) = 49.5 and 3(IQR) = 3(33) = 99. Observations that are further than 49.5 below the lower quartile (i.e., 359–49.5 = 309.5 or less) or more than 49.5 units above the upper quartile (greater than 392+49.5 = 441.5) are classified as 'mild' outliers. 'Extreme' outliers would fall 99 or more units below the lower, or above the upper, quartile. Since the minimum and maximum observations in the data are 325 and 424, we conclude that there are no mild outliers in this data (and therefore, no 'extreme' outliers either).

c. A boxplot (created by Minitab) of this data appears below. There is a slight positive skew to the data, but it is not far from being symmetric. The variation, however, seems large (the spread 424–325 = 99 is a large percentage of the median/typical value)

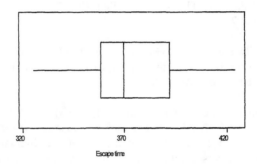

Escape time

d. Not until the value x = 424 is lowered below the upper quartile value of 392 would there be any change in the value of the upper quartile. That is, the value x = 424 could not be decreased by more than 424–392 = 32 units.

57.

a. 1.5(IQR) = 1.5(216.8–196.0) = 31.2 and 3(IQR) = 3(216.8–196.0) = 62.4. Mild outliers: observations below 196–31.2 = 164.6 or above 216.8+31.2 = 248. Extreme outliers: observations below 196–62.4 = 133.6 or above 216.8+62.4 = 279.2. Of the observations given, 125.8 is an extreme outlier and 250.2 is a mild outlier.

b. A boxplot of this data appears below. There is a bit of positive skew to the data but, except for the two outliers identified in part (a), the variation in the data is relatively small.

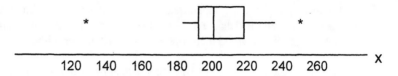

59.

a. ED: median = .4 (the 14th value in the *sorted* list of data). The lower quartile (median of the lower half of the data, including the median, since n is odd) is (.1+.1)/2 = .1. The upper quartile is (2.7+2.8)/2 = 2.75. Therefore, IQR = 2.75 – .1 = 2.65.

Non–ED: median = (1.5+1.7)/2 = 1.6. The lower quartile (median of the lower 25 observations) is .3; the upper quartile (median of the upper half of the data) is 7.9. Therefore, IQR = 7.9 – .3 = 7.6.

b. ED: mild outliers are less than .1 – 1.5(2.65) = –3.875 or greater than 2.75 + 1.5(2.65) = 6.725. Extreme outliers are less than .1 – 3(2.65) = –7.85 or greater than 2.75 + 3(2.65) = 10.7. So, the two largest observations (11.7, 21.0) are extreme outliers and the next two largest values (8.9, 9.2) are mild outliers. There are no outliers at the lower end of the data.

Non–ED: mild outliers are less than .3 – 1.5(7.6) = –11.1 or greater than 7.9 + 1.5(7.6) = 19.3. Note that there are no mild outliers in the data, hence there can not be any extreme outliers either.

c.   A comparative boxplot appears below. The outliers in the ED data are clearly
visible. There is noticeable positive skewness in both samples; the Non–Ed data
has more variability then the Ed data; the typical values of the ED data tend to be
smaller than those for the Non–ED data.

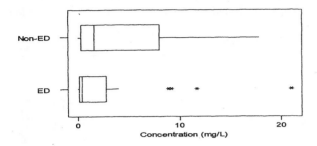

61.   Outliers occur in the 6 a.m. data. The distributions at the other times are fairly
symmetric. Variability and the 'typical' values in the data increase a little at the 12
noon and 2 p.m. times.

Supplementary Exercises

63. A stem plot and a box plot appear below. We see that the distribution of solar radiation measurements is negatively skewed with no outliers. Summary statistics: $\bar{x} = 9.96$, $\tilde{x} = 10.6$, $s = 1.76$, and $f_s = 2.3$ (lower fourth = 8.85, upper fourth = 11.15).

6	34
7	17
8	4589
9	1
10	12667789
11	122499
12	2
13	1

stem=ones
leaf=tenths

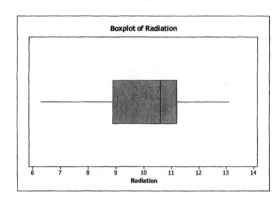

65.

a. The histogram appears below. A representative value for this data would be $x = 90$. The histogram is reasonably symmetric, unimodal, and somewhat bell–shaped. The variation in the data is not small since the spread of the data (99–81 = 18) constitutes about 20% of the typical value of 90.

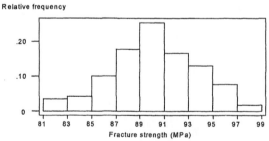

b. The proportion of the observations that are at least 85 is $1 - (6+7)/169 = .9231$. The proportion less than 95 is $1 - (13+3)/169 = .9053$.

c. $x = 90$ is the midpoint of the class 89–<91, which contains 43 observations (a relative frequency of $43/169 = .2544$. Therefore about half of this frequency, .1272, should be added to the relative frequencies for the classes to the left of $x = 90$. That is, the approximate proportion of observations that are less than 90 is $.0355 + .0414 + .1006 + .1775 + .1272 = .4822$.

67.

   a. Aortic root diameters for males have mean 3.64 cm, median 3.70 cm, standard deviation 0.269 cm, and fourth spread 0.40. The corresponding values for females are $\bar{x} = 3.28$ cm, $\tilde{x} = 3.15$ cm, $s = 0.478$ cm, and $f_s = 0.50$ cm. Aortic root diameters are typically (though not universally) somewhat smaller for females than for males, and females show more variability. The distribution for males is negatively skewed, while the distribution for females is positively skewed (see graphs below).

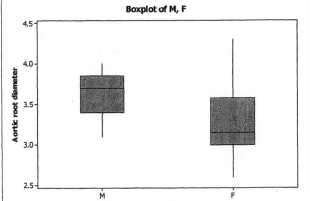

   b. For females ($n = 10$), the 10% trimmed mean is the average of the middle 8 observations: $\bar{x}_{tr(10)} = 3.24$ cm. For males ($n = 13$), the 1/13 trimmed mean is $40.2/11 = 3.6545$, and the 2/13 trimmed mean is $32.8/9 = 3.6444$. Interpolating, the 10% trimmed mean is $\bar{x}_{tr(10)} = 0.7(3.6545) + 0.3(3.6444) = 3.65$ cm. (10% is three–tenths of the way from 1/13 to 2/13.)

69.

   a.

$$\bar{y} = \frac{\sum y_i}{n} = \frac{\sum (ax_i + b)}{n} = \frac{a\sum x_i + b}{n} = a\bar{x} + b.$$

$$s_y^2 = \frac{\sum (y_i - \bar{y})^2}{n-1} = \frac{\sum (ax_i + b - (a\bar{x} + b))^2}{n-1} = \frac{\sum (ax_i - a\bar{x})^2}{n-1}$$

$$= \frac{a^2 \sum (x_i - \bar{x})^2}{n-1} = a^2 s_x^2.$$

b.

$x=°C, y=°F$

$$\bar{y} = \frac{9}{5}(87.3) + 32 = 189.14$$

$$s_y = \sqrt{s_y^2} = \sqrt{\left(\frac{9}{5}\right)^2 (1.04)^2} = \sqrt{3.5044} = 1.872$$

71.

    a. The mean, median, and trimmed mean are virtually identical, which suggests symmetry. If there are outliers, they are balanced. The range of values is only 25.5, but half of the values are between 132.95 and 138.25.

    b.

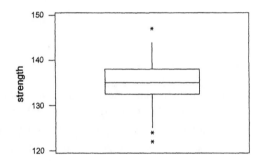

The boxplot also displays the symmetry, and adds a visual of the outliers, two on the lower end, and one on the upper.

73.

```
0.7 8              stem=tenths
0.8 11556          leaf=hundredths
0.9 2233335566
1.0 0566
```

$\bar{x} = .9255, s = .0809, \tilde{x} = .93$

$lowerfourth = .855, upperfourth = .96$

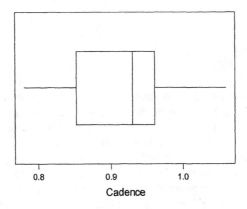

The data appears to be a bit skewed toward smaller values (negatively skewed). There are no outliers. The mean and the median are close in value.

75.

a. The median is the same (371) in each plot and all three data sets are very symmetric. In addition, all three have the same minimum value (350) and same maximum value (392). Moreover, all three data sets have the same lower (364) and upper quartiles (378). So, all three boxplots will be *identical*.

b. A comparative dotplot is shown below. These graphs show that there are differences in the variability of the three data sets. They also show differences in the way the values are distributed in the three data sets.

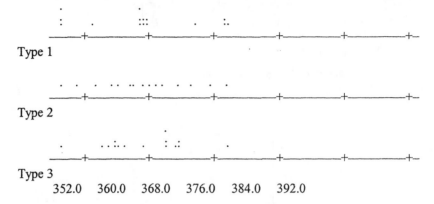

Type 1

Type 2

Type 3

352.0　　360.0　　368.0　　376.0　　384.0　　392.0

c. The boxplot in (a) is not capable of detecting the differences among the data sets. The primary reason is that boxplots give up some detail in describing data because they use only 5 summary numbers for comparing data sets. Note: The definition of lower and upper quartile used in this text is slightly different than the one used by some other authors (and software packages). Technically speaking, the median of the lower half of the data is not really the first quartile, although it is generally *very close*. Instead, the medians of the lower and upper halves of the data are often called the **lower** and **upper hinges.** Our boxplots use the lower and upper hinges to define the spread of the middle 50% of the data, but other authors sometimes use the *actual* quartiles for this purpose. The difference is usually very slight, usually unnoticeable, but not always. For example in the data sets of this exercise, a comparative boxplot based on the actual quartiles (as computed by Minitab) is shown below. The graph shows substantially the same type of information as those described in (a) except the graphs based on quartiles are able to detect the slight differences in variation between the three data sets.

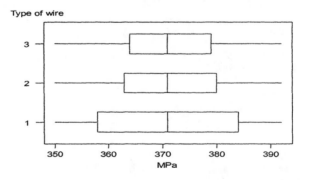

77.

a.

```
 0|2355566777888
 1|0000135555
 2|00257
 3|0033
 4|0057
 5|044
 6|                    stem: ones
 7|05                  leaf: tenths
 8|8
 9|0
10|3
HI|22.0 24.5
```

b.

Interval	Frequency	Rel. Freq.	Density
0 –< 2	23	.500	.250
2 –< 4	9	.196	.098
4 –< 6	7	.152	.076
6 –< 10	4	.087	.022
10 –< 20	1	.022	.002
20 –< 30	2	.043	.004

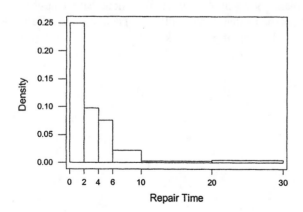

79.

a. $\sum_{i=1}^{n+1} x_i = \sum_{i=1}^{n} x_i + x_{n+1} = n\bar{x}_n + x_{n+1}, \ so \ \bar{x}_{n+1} = \dfrac{[n\bar{x}_n + x_{n+1}]}{(n+1)}$

b.

$$ns_{n+1}^2 = \sum_{i=1}^{n+1}(x_i - \bar{x}_{n+1})^2 = \sum_{i=1}^{n+1} x_i^2 - (n+1)\bar{x}_{n+1}^2$$

$$= \sum_{i=1}^{n} x_i^2 - n\bar{x}_n^2 + x_{n+1}^2 + n\bar{x}_n^2 - (n+1)\bar{x}_{n+1}^2$$

$$= (n-1)s_n^2 + \left\{ x_{n+1}^2 + n\bar{x}_n^2 - (n+1)\bar{x}_{n+1}^2 \right\}$$

When the expression for $\bar{x}_{n+1}$ from **a** is substituted, the expression in braces simplifies to the following, as desired: $\dfrac{n(x_{n+1} - \bar{x}_n)^2}{(n+1)}$

81. Assuming that the histogram is unimodal, then there is evidence of positive skewness in the data since the median lies to the left of the mean (for a symmetric distribution, the mean and median would coincide). For more evidence of skewness, compare the distances of the 5th and 95th percentiles from the median: median – 5th percentile = 500 – 400 = 100, while 95th percentile –median = 720 – 500 = 220. Thus, the largest 5% of the values (above the 95th percentile) are further from the median than are the lowest 5%. The same skewness is evident when comparing the 10th and 90th percentiles to the median: median – 10th percentile = 500 – 430 = 70, while 90th percentile –median = 640 – 500 = 140. Finally, note that the largest value (925) is much further from the median (925 – 500 = 425) than is the smallest value (500 – 220 = 280), again an indication of positive skewness.

83.

    a.   When there is perfect symmetry, the smallest observation y_1 and the largest observation y_n will be equidistant from the median, so $y_n - \tilde{x} = \tilde{x} - y_1$. Similarly, the second smallest and second largest will be equidistant from the median, so $y_{n-1} - \tilde{x} = \tilde{x} - y_2$, and so on. Thus, the first and second numbers in each pair will be equal, so that each point in the plot will fall exactly on the 45 degree line. When the data is positively skewed, y_n will be much further from the median than is y_1, so $y_n - \tilde{x}$ will considerably exceed $\tilde{x} - y_1$ and the point $(y_n - \tilde{x}, \tilde{x} - y_1)$ will fall considerably below the 45 degree line. A similar comment aplies to other points in the plot.

    b.   The first point in the plot is $(2745.6 - 221.6, 221.6\,0 - 4.1) = (2524.0, 217.5)$. The others are: (1476.2, 213.9), (1434.4, 204.1), (756.4, 190.2), (481.8, 188.9), (267.5, 181.0), (208.4, 129.2), (112.5, 106.3), (81.2, 103.3), (53.1, 102.6), (53.1, 92.0), (33.4, 23.0), and (20.9, 20.9). The first number in each of the first seven pairs greatly exceed the second number, so each point falls well below the 45 degree line. A substantial positive skew (stretched upper tail) is indicated.

CHAPTER 2

Section 2.1

1.

    a. S = {1324, 1342, 1423, 1432, 2314, 2341, 2413, 2431, 3124, 3142, 4123, 4132, 3214, 3241, 4213, 4231}

    b. Event A contains the outcomes where 1 is first in the list:
A = {1324, 1342, 1423, 1432}

    c. Event B contains the outcomes where 2 is first or second:
B = {2314, 2341, 2413, 2431, 3214, 3241, 4213, 4231}

    d. The compound event A∪B contains the outcomes in A or B or both:
A∪B = {1324, 1342, 1423, 1432, 2314, 2341, 2413, 2431, 3214, 3241, 4213, 4231}
A∩B = ∅, since 1 and 2 can't both get into the championship game
A′ = S – A = {2314, 2341, 2413, 2431, 3124, 3142, 4123, 4132, 3214, 3241, 4213, 4231}

3.

    a. Event A = { SSF, SFS, FSS }

    b. Event B = { SSS, SSF, SFS, FSS }

    c. For Event C, the system must have component 1 working (S in the first position), then at least one of the other two components must work (at least one S in the 2nd and 3rd positions: Event C = { SSS, SSF, SFS }

    d. Event C′ = { SFF, FSS, FSF, FFS, FFF }
Event A∪C = { SSS, SSF, SFS, FSS }
Event A∩C = { SSF, SFS }
Event B∪C = { SSS, SSF, SFS, FSS }
Event B∩C = { SSS SSF, SFS }

5.

    a.

Outcome Number	Outcome	Outcome Number	Outcome
1	111	15	223
2	112	16	231
3	113	17	232
4	121	18	233
5	122	19	311
6	123	20	312
7	131	21	313
8	132	22	321
9	133	23	322
10	211	24	323
11	212	25	331
12	213	26	332
13	221	27	333
14	222		

    b.   Outcome Numbers 1, 14, 27

    c.   Outcome Numbers 6, 8, 12, 16, 20, 22

    d.   Outcome Numbers 1, 3, 7, 9, 19, 21, 25, 27

7.

    a.   S = {BBBAAAA, BBABAAA, BBAABAA, BBAAABA, BBAAAAB, BABBAAA, BABABAA, BABAABA, BABAAAB, BAABBAA, BAABABA, BAABAAB, BAAABBA, BAAABAB, BAAAABB, ABBBAAA, ABBABAA, ABBAABA, ABBAAAB, ABABBAA, ABABABA, ABABAAB, ABAABBA, ABAABAB, ABAAABB, AABBBAA, AABBABA, AABBAAB, AABABBA, AABABAB, AABAABB, AAABBBA, AAABBAB, AAABABB, AAAABBB}

    b.   {AAAABBB, AAABABB, AAABBAB, AABAABB, AABABAB}

9.

    a. In the diagram on the left, the shaded area is (A∪B)′. On the right, the shaded area is A′, the striped area is B′, and the intersection A′ ∩ B′ occurs where there is BOTH shading and stripes. These two diagrams display the same area.

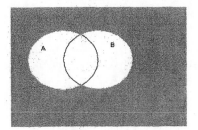 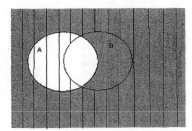

    b. In the diagram below, the shaded area represents (A∩B)′. Using the diagram on the right above, the union of A′ and B′ is represented by the areas that have either shading or stripes or both. Both of the diagrams display the same area.

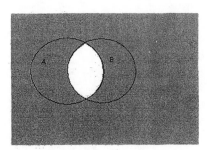 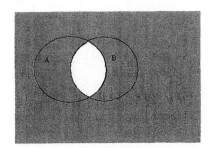

Section 2.2

11.

    a. .07

    b. .15 + .10 + .05 = .30

    c. Let event A = selected customer owns stocks. Then the probability that a selected customer does not own a stock can be represented by P(A′) = 1 − P(A) = 1 − (.18 + .25) = 1 − .43 = .57. This could also have been done easily by adding the probabilities of the funds that are not stocks.

13.

a. awarded either #1 or #2 (or both):

$P(A_1 \cup A_2) = P(A_1) + P(A_2) - P(A_1 \cap A_2) = .22 + .25 - .11 = .36$

b. awarded neither #1 or #2:

$P(A_1' \cap A_2') = P[(A_1 \cup A_2)'] = 1 - P(A_1 \cup A_2) = 1 - .36 = .64$

c. awarded at least one of #1, #2, #3:

$P(A_1 \cup A_2 \cup A_3) = P(A_1) + P(A_2) + P(A_3) - P(A_1 \cap A_2) - P(A_1 \cap A_3) -$
$P(A_2 \cap A_3) + P(A_1 \cap A_2 \cap A_3)$
$= .22 + .25 + .28 - .11 - .05 - .07 + .01 = .53$

d. awarded none of the three projects:

$P(A_1' \cap A_2' \cap A_3') = 1 - P(\text{awarded at least one}) = 1 - .53 = .47.$

e. awarded #3 but neither #1 nor #2:

$P(A_1' \cap A_2' \cap A_3) = P(A_3) - P(A_1 \cap A_3) - P(A_2 \cap A_3)$
$+ P(A_1 \cap A_2 \cap A_3)$
$= .28 - .05 - .07 + .01 \quad = .17$

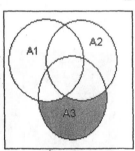

Chapter 2: Probability

f. either (neither #1 nor #2) or #3: $P[(A_1' \cap A_2') \cup A_3]$
= P(shaded region) = P(awarded none) + $P(A_3)$ = .47 + .28 = .75. Alternatively, answers to **a – f** can be obtained from probabilities on the Venn diagram on the right

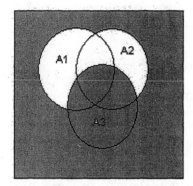 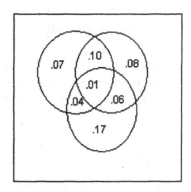

15.

a. Let event E be the event that at most one purchases an electric dryer. Then E′ is the event that at least two purchase electric dryers.
$P(E') = 1 - P(E) = 1 - .428 = .572$

b. Let event A be the event that all five purchase gas. Let event B be the event that all five purchase electric. All other possible outcomes are those in which at least one of each type is purchased. Thus, the desired probability =
$1 - P(A) - P(B) = 1 - .116 - .005 = .879$

17.

a. The probabilities do not add to 1 because there are other software packages besides SPSS and SAS for which requests could be made.

b. $P(A') = 1 - P(A) = 1 - .30 = .70$

c. $P(A \cup B) = P(A) + P(B) = .30 + .50 = .80$
(since A and B are mutually exclusive events)

d. $P(A' \cap B') = P[(A \cup B)']$ (De Morgan's law)
$= 1 - P(A \cup B)$
$= 1 - .80 = .20$

19.   Let event A be that the selected joint was found defective by inspector A. P(A) = $\frac{724}{10,000}$. Let event B be analogous for inspector B. P(B) = $\frac{751}{10,000}$. Compound event A∪B is the event that the selected joint was found defective by at least one of the two inspectors. P(A∪B) = $\frac{1159}{10,000}$.

a. The desired event is (A∪B)′, so we use the complement rule:
    P(A∪B)′ = 1 – P(A∪B) = 1 – $\frac{1159}{10,000}$ = $\frac{8841}{10,000}$ = .8841

b. The desired event is B ∩ A′. P(B ∩ A′) = P(B) – P(A ∩ B).
    P(A ∩ B) = P(A) + P(B) – P(A∪B), = .0724 + .0751 – .1159 = .0316
    So P(B ∩ A′) = P(B) – P(A ∩ B) = .0751 – .0316 = .0435

21.

a. P({M,H}) = .10

b. P(low auto) = P[{(L,N}, (L,L), (L,M), (L,H)}] = .04 + .06 + .05 + .03 = .18
    Following a similar pattern, P(low homeowner's) = .06 + .10 + .03 = .19

c. P(same deductible for both) = P[{ LL, MM, HH }] = .06 + .20 + .15 = .41

d. P(deductibles are different) = 1 – P(same deductibles) = 1 – .41 = .59

e. P(at least one low deductible) = P[{ LN, LL, LM, LH, ML, HL }]
                            = .04 + .06 + .05 + .03 + .10 + .03 = .31

f. P(neither low) = 1 – P(at least one low) = 1 – .31 = .69

23. Assume that the computers are numbered 1 – 6 as described. Also assume that computers 1 and 2 are the laptops. Possible outcomes are (1,2) (1,3) (1,4) (1,5) (1,6) (2,3) (2,4) (2,5) (2,6) (3,4) (3,5) (3,6) (4,5) (4,6) and (5,6).

    a. P(both are laptops) = P[{ (1,2)}] = $\frac{1}{15}$ = .067

    b. P(both are desktops) = P[{(3,4) (3,5) (3,6) (4,5) (4,6) (5,6)}] = $\frac{6}{15}$ = .40

    c. P(at least one desktop) = 1 – P(no desktops)
       = 1 – P(both are laptops)   = 1 – .067 = .933

    d. P(at least one of each type) = 1 – P(both are the same)
       = 1 – P(both laptops) – P(both desktops) = 1 – .067 – .40 = .533

25. $P(A \cap B) = P(A) + P(B) - P(A \cup B) = .65$
$P(A \cap C) = .55, \ P(B \cap C) = .60$
$P(A \cap B \cap C) = P(A \cup B \cup C) - P(A) - P(B) - P(C)$
$$+ P(A \cap B) + P(A \cap C) + P(B \cap C)$$
$$= .98 - .7 - .8 - .75 + .65 + .55 + .60$$
$$= .53$$

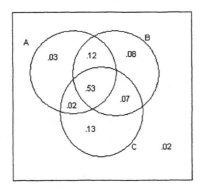

    a. $P(A \cup B \cup C) = .98$, as given.

    b. P(none selected) = 1 – P(A ∪ B ∪ C) = 1 – .98 = .02

    c. P(only automatic transmission selected) = .03 from the Venn Diagram

    d. P(exactly one of the three) = .03 + .08 + .13 = .24

27.   Outcomes: (A,B) (A,C$_1$) (A,C$_2$) (A,F) (B,A) (B,C$_1$) (B,C$_2$) (B,F)
                      (C$_1$,A) (C$_1$,B) (C$_1$,C$_2$) (C$_1$,F) (C$_2$,A) (C$_2$,B) (C$_2$,C$_1$) (C$_2$,F)
                      (F,A) (F,B) (F,C$_1$) (F,C$_2$)

   a.   P[(A,B) or (B,A)] = $\frac{2}{20} = \frac{1}{10} = .1$

   b.   P(at least one C) = $\frac{14}{20} = \frac{7}{10} = .7$

   c.   P(at least 15 years) = 1 – P(at most 14 years)
        = 1 – P[(3,6) or (6,3) or (3,7) or (7,3) or (3,10) or (10,3) or (6,7) or (7,6)]
        = $1 - \frac{8}{20} = 1 - .4 = .6$

Section 2.3

29.
   a. $(26)(26) = 676$; $(36)(36) = 1296$

   b. $(26)^3 = 17{,}576$; $(36)^3 = 46{,}656$

   c. $(26)^3 = 456{,}976$; $(36)^4 = 1{,}679{,}616$

   d. $1 - 97{,}786/(36)^4 = .942$

31.
   a. $(n_1)(n_2) = (9)(27) = 243$

   b. $(n_1)(n_2)(n_3) = (9)(27)(15) = 3645$, so such a policy could be carried out for 3645 successive nights, or approximately 10 years, without repeating exactly the same program.

33.

    a. $9! = 362,880$

    b.   9! Starting line-ups and 9! batting orders $\rightarrow$ $(9!)(9!) = 131,681,894,400$

    c. $\binom{5}{3} \times \binom{10}{6} = (10)(210) = 2,100$

35.

    a. $\binom{20}{6} = 38,760$. P(all from day shift) $= \dfrac{\binom{20}{6}\binom{25}{0}}{\binom{45}{6}} = \dfrac{38,760}{8,145,060} = .0048$

    b.   P(all from same shift) $= \dfrac{\binom{20}{6}\binom{25}{0}}{\binom{45}{6}} + \dfrac{\binom{15}{6}\binom{30}{0}}{\binom{45}{6}} + \dfrac{\binom{10}{6}\binom{35}{0}}{\binom{45}{6}} = .0048 + .0006 +$

        $.0000 = .0054$

    c.   P(at least two shifts represented) $= 1 - $ P(all from same shift) $= 1 - .0054 = .9946$

    d.   Let $A_1 = $ day shift unrepresented, $A_2 = $ swing shift unrepresented, and $A_3 = $ graveyard shift unrepresented. Then we want $P(A_1 \cup A_2 \cup A_3)$.

        $P(A_1) = $ P(day unrepresented) $= $ P(all from swing and graveyard) $= \dfrac{\binom{25}{6}}{\binom{45}{6}}$.

        Similarly,

$$P(A_2) = \frac{\binom{30}{6}}{\binom{45}{6}} \text{ and } P(A_3) = \frac{\binom{35}{6}}{\binom{45}{6}}. \text{ Next, } P(A_1 \cap A_2) = P(\text{all from graveyard}) =$$

$$\frac{\binom{10}{6}}{\binom{45}{6}}. \text{ Similarly, } P(A_1 \cap A_3) = \frac{\binom{15}{6}}{\binom{45}{6}} \text{ and } P(A_2 \cap A_3) = \frac{\binom{20}{6}}{\binom{45}{6}}.$$

Finally, $P(A_1 \cap A_2 \cap A_3) = 0$, since at least one shift must be represented.

$$\text{So, } P(A_1 \cup A_2 \cup A_3) = \frac{\binom{25}{6}}{\binom{45}{6}} + \frac{\binom{30}{6}}{\binom{45}{6}} + \frac{\binom{35}{6}}{\binom{45}{6}} - \frac{\binom{10}{6}}{\binom{45}{6}} - \frac{\binom{15}{6}}{\binom{45}{6}} - \frac{\binom{20}{6}}{\binom{45}{6}} + 0 =$$

.2885.

37.

    a. $n_1 = 3$, $n_2 = 4$, $n_3 = 5$, so $n_1 \times n_2 \times n_3 = 60$ runs

    b. $n_1 = 1$, (just one temperature), $n_2 = 2$, $n_3 = 5$ implies that there are 10 such runs.

    c.   There are $\binom{60}{5}$ ways to select the 5 runs. Each catalyst is used in 12 different runs, so the number of ways of selecting one run from each of these 5 groups is 12^5. Thus the desired probability is $\dfrac{12^5}{\binom{60}{5}} = .0456$.

39.

a. We want to choose all of the 5 cordless, and 5 of the 10 others, to be among the first 10 serviced, so the desired probability is $\dfrac{\dbinom{5}{5}\dbinom{10}{5}}{\dbinom{15}{10}} = \dfrac{252}{3003} = .0839$

b. Isolating one group, say the cordless phones, we want the other two groups represented in the last 5 serviced. So we choose 5 of the 10 others, except that we don't want to include the outcomes where the last five are all the same.

So we have $\dfrac{\dbinom{10}{5} - 2}{\dbinom{15}{5}}$. But we have three groups of phones, so the desired

probability is $\dfrac{3 \cdot \left[\dbinom{10}{5} - 2\right]}{\dbinom{15}{5}} = \dfrac{3(250)}{3003} = .2498$.

c. We want to choose 2 of the 5 cordless, 2 of the 5 cellular, and 2 of the corded phones: $\dfrac{\dbinom{5}{2}\dbinom{5}{2}\dbinom{5}{2}}{\dbinom{15}{6}} = \dfrac{1000}{5005} = .1998$

41.

a. P(at least one F among 1ˢᵗ 3) = 1 − P(no F's among 1ˢᵗ 3)

$$= 1 - \frac{4 \times 3 \times 2}{8 \times 7 \times 6} = 1 - \frac{24}{336} = 1 - .0714 = .9286$$

An alternative method to calculate P(no F's among 1ˢᵗ 3)
would be to choose none of the females and 3 of the 4 males, as follows:

$$\frac{\binom{4}{0}\binom{4}{3}}{\binom{8}{3}} = \frac{4}{56} = .0714 \text{, obviously producing the same result.}$$

b. P(all F's among 1ˢᵗ 5) = $\dfrac{\binom{4}{4}\binom{4}{1}}{\binom{8}{5}} = \dfrac{4}{56} = .0714$

c. P(orderings are different) = 1 − P(orderings are the same for both semesters)
There are 8! possible orderings for the second semester, only one of which would
match the first semester (whatever ordering that was). So, P(orderings are
different) = 1 − 1/8! = .999975.

43.   # of 10 high straights = 4×4×4×4×4 (4 – 10's, 4 – 9's , etc)

$$P(10 \text{ high straight}) = \frac{4^5}{\binom{52}{5}} = \frac{1024}{2{,}598{,}960} = .000394$$

$$P(\text{straight}) = 10 \times \frac{4^5}{\binom{52}{5}} = .003940 \text{ (Multiply by 10 because there are 10 different}$$

card values that could be high: Ace, King, etc.) There are only 40 straight flushes (10 in each suit), so

$$P(\text{straight flush}) = \frac{40}{\binom{52}{5}} = .00001539$$

Section 2.4

45.

    a.   P(A) = .106 + .141 + .200 = .447, P(C) =.215 + .200 + .065 + .020 = .500,
          $P(A \cap C) = .200$

    b. $P(A|C) = \dfrac{P(A \cap C)}{P(C)} = \dfrac{.200}{.500} = .400$. If we know that the individual came from
        ethnic group 3, the probability that he has type A blood is .40. P(C|A) =
        $\dfrac{P(A \cap C)}{P(A)} = \dfrac{.200}{.447} = .447$. If a person has type A blood, the probability that he is
        from ethnic group 3 is .447

    c.   Define event D = {ethnic group 1 selected}. We are asked for P(D|B′) =
        $\dfrac{P(D \cap B')}{P(B')} = \dfrac{.192}{.909} = .211$. P(D∩B′)=.082 + .106 + .004 = .192, P(B′) = 1 – P(B)
        = 1 – [.008 + .018 + .065] = .909

47.

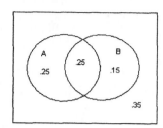

a. $P(B \mid A) = \dfrac{P(A \cap B)}{P(A)} = \dfrac{.25}{.50} = .50$

b. $P(B' \mid A) = \dfrac{P(A \cap B')}{P(A)} = \dfrac{.25}{.50} = .50$

c. $P(A \mid B) = \dfrac{P(A \cap B)}{P(B)} = \dfrac{.25}{.40} = .6125$

d. $P(A' \mid B) = \dfrac{P(A' \cap B)}{P(B)} = \dfrac{.15}{.40} = .3875$

e. $P(A \mid A \cup B) = \dfrac{P[A \cap (A \cup B)]}{P(A \cup B)} = \dfrac{.50}{.65} = .7692$

49. The first desired probability is P(both bulbs are 75 watt | at least one is 75 watt).

P(at least one is 75 watt) = 1 − P(none are 75 watt)

$$= 1 - \frac{\binom{9}{2}}{\binom{15}{2}} = 1 - \frac{36}{105} = \frac{69}{105}.$$

Notice that P[(both are 75 watt)∩(at least one is 75 watt)]

$$= \text{P(both are 75 watt)} = \frac{\binom{6}{2}}{\binom{15}{2}} = \frac{15}{105}.$$

So P(both bulbs are 75 watt | at least one is 75 watt) $= \dfrac{\frac{15}{105}}{\frac{69}{105}} = \dfrac{15}{69} = .2174$

Second, we want P(same rating | at least one NOT 75 watt).

P(at least one NOT 75 watt) = 1 − P(both are 75 watt)

$$= 1 - \frac{15}{105} = \frac{90}{105}.$$

Now, P[(same rating)∩(at least one not 75 watt)] = P(both 40 watt or both 60 watt).

$$\text{P(both 40 watt or both 60 watt)} = \frac{\binom{4}{2}+\binom{5}{2}}{\binom{15}{2}} = \frac{16}{105}$$

Now, the desired conditional probability is $\dfrac{\frac{16}{105}}{\frac{90}{105}} = \dfrac{16}{90} = .1778$

51.

    a. $P(\text{R from }1^{st} \cap \text{R from }2^{nd}) = P(\text{R from }2^{nd} \mid \text{R from }1^{st}) \bullet P(\text{R from }1^{st})$

$$= \frac{8}{11} \times \frac{6}{10} = .436$$

    b. P(same numbers)   = P(both selected balls are the same color)

$$= \text{P(both red) + P(both green)} = .436 + \frac{4}{11} \times \frac{4}{10} = .581$$

53. $P(B \mid A) = \dfrac{P(A \cap B)}{P(A)} = \dfrac{P(B)}{P(A)} = \dfrac{.05}{.60} = .0833$ (since B is contained in A, $A \cap B = B$)

55.   Let A = {carries Lyme disease} and B = {carries HGE}. We are told P(A) = .16, P(B) = .10, and $P(A \cap B \mid A \cup B) = .10$. From this last statement and the fact that $A \cap B$ is contained in $A \cup B$,

$.10 = \dfrac{P(A \cap B)}{P(A \cup B)} \rightarrow P(A \cap B) = .10P(A \cup B) = .10[P(A) + P(B) - P(A \cap B)] = .10[.10 +$

$.16 - P(A \cap B)] \rightarrow 1.1P(A \cap B) = .026 \rightarrow P(A \cap B) = .02364.$

Finally, the desired probability is P(A | B) = $\dfrac{P(A \cap B)}{P(B)} = \dfrac{.02364}{.10} = .2364.$

57.   P(B|A) > P(B)

    $\leftrightarrow$ P(B|A) + P(B'|A) > P(B) + P(B'|A)

    $\leftrightarrow$ 1 > P(B) + P(B'|A) by Exercise 56 (with the letters switched)

    $\leftrightarrow$ 1 − P(B) > P(B'|A)

    $\leftrightarrow$ P(B') > P(B'|A), QED

59.

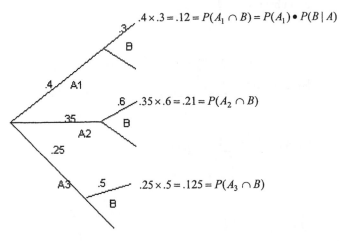

$.4 \times .3 = .12 = P(A_1 \cap B) = P(A_1) \bullet P(B \mid A)$

$.35 \times .6 = .21 = P(A_2 \cap B)$

$.25 \times .5 = .125 = P(A_3 \cap B)$

a. $P(A_2 \cap B) = .21$

b. $P(B) = P(A_1 \cap B) + P(A_2 \cap B) + P(A_3 \cap B) = .455$

c. $P(A_1|B) = \dfrac{P(A_1 \cap B)}{P(B)} = \dfrac{.12}{.455} = .264$, $P(A_2|B) = \dfrac{.21}{.455} = .462$, $P(A_3|B) = 1 - .264 -$

$.462 = .274$

61. $P(0 \text{ def in sample} \mid 0 \text{ def in batch}) = 1$

$P(0 \text{ def in sample} \mid 1 \text{ def in batch}) = \dfrac{\binom{9}{2}}{\binom{10}{2}} = .800$

$P(1 \text{ def in sample} \mid 1 \text{ def in batch}) = \dfrac{\binom{9}{1}}{\binom{10}{2}} = .200$

$$P(0 \text{ def in sample} \mid 2 \text{ def in batch}) = \frac{\binom{8}{2}}{\binom{10}{2}} = .622$$

$$P(1 \text{ def in sample} \mid 2 \text{ def in batch}) = \frac{\binom{2}{1}\binom{8}{1}}{\binom{10}{2}} = .356$$

$$P(2 \text{ def in sample} \mid 2 \text{ def in batch}) = \frac{1}{\binom{10}{2}} = .022$$

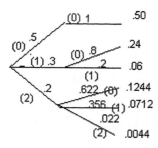

a. $P(0 \text{ def in batch} \mid 0 \text{ def in sample}) = \dfrac{.5}{.5 + .24 + .1244} = .578$

$P(1 \text{ def in batch} \mid 0 \text{ def in sample}) = \dfrac{.24}{.5 + .24 + .1244} = .278$

$P(2 \text{ def in batch} \mid 0 \text{ def in sample}) = \dfrac{.1244}{.5 + .24 + .1244} = .144$

b. $P(0 \text{ def in batch} \mid 1 \text{ def in sample}) = 0$

$P(1 \text{ def in batch} \mid 1 \text{ def in sample}) = \dfrac{.06}{.06 + .0712} = .457$

$P(2 \text{ def in batch} \mid 1 \text{ def in sample}) = \dfrac{.0712}{.06 + .0712} = .543$

63.

a.

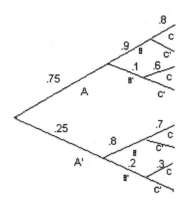

b. $P(A \cap B \cap C) = .75 \times .9 \times .8 = .5400$

c. $P(B \cap C) = P(A \cap B \cap C) + P(A' \cap B \cap C) = .5400 + .25 \times .8 \times .7 = .6800$

d. $P(C) = P(A \cap B \cap C) + P(A' \cap B \cap C) + P(A \cap B' \cap C) + P(A' \cap B' \cap C)$
    $= .54 + .045 + .14 + .015 = .74$

e. $P(A|B \cap C) = \dfrac{P(A \cap B \cap C)}{P(B \cap C)} = \dfrac{.54}{.68} = .7941$

65.

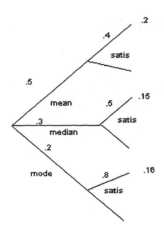

$P(\text{satis}) = .51$

$P(\text{mean} \mid \text{satis}) = \dfrac{.2}{.51} = .3922$

$P(\text{median} \mid \text{satis}) = .2941$

$P(\text{mode} \mid \text{satis}) = .3137$

So Mean (and not Mode!) is the most
    likely author, while Median is least.

67. Let T denote the event that a randomly selected person is, in fact, a terrorist. Apply Bayes' Theorem, using $P(T) = 1,000/300,000,000 = .0000033$:

$$P(T|+) = \frac{P(T)P(+|T)}{P(T)P(+|T) + P(T')P(+|T')} = \frac{(.0000033)(.99)}{(.0000033)(.99) + (1 - .0000033)(1 - .999)} =$$

.003289. That is to say, roughly 0.3% of all people "flagged" as terrorists would be actua terrorists in this scenario.

69.

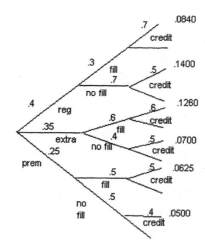

a. $P(U \cap F \cap Cr) = .1260$

d. $P(F \cap Cr) = .0840 + .1260 + .0625 = .2725$

b. $P(Pr \cap NF \cap Cr) = .05$

e. $P(Cr) = .5325$

c. $P(Pr \cap Cr) = .0625 + .05$
    $= .1125$

f. $P(PR | Cr) =$
    $\dfrac{P(Pr \cap Cr)}{P(Cr)} = \dfrac{.1125}{.5325} = .2113$

Section 2.5

71.

    a. Since the events are independent, then A' and B' are independent, too. (See paragraph below equation 2.7.) $P(B'|A') = . \quad P(B') = 1 - .7 = .3$

    b. $P(A \cup B) = P(A) + P(B) - P(A) \cdot P(B) = .4 + .7 - (.4)(.7) = .82$

    c. $P(AB' | A \cup B) = \dfrac{P(AB' \cap (A \cup B))}{P(A \cup B)} = \dfrac{P(AB')}{P(A \cup B)} = \dfrac{.12}{.82} = .146$

73. $P(A' \cap B) = P(B) - P(A \cap B) = P(B) - P(A)P(B) = [1 - P(A)]P(B) = P(A')P(B)$.

    Alternatively, $P(A' | B) = \dfrac{P(A' \cap B)}{P(B)} = \dfrac{P(B) - P(A \cap B)}{P(B)}$

    $= \dfrac{P(B) - P(A) \cdot P(B)}{P(B)} = 1 - P(A) = P(A')$.

75.   Let event E be the event that an error was signaled incorrectly. We want P(at least one signaled incorrectly) $= P(E_1 \cup E_2 \cup \ldots \cup E_{10}) = 1 - P(E_1' \cap E_2' \cap \ldots \cap E_{10}')$. $P(E') = 1 - .05 = .95$. For 10 independent points, $P(E_1' \cap E_2' \cap \ldots \cap E_{10}') = P(E_1')P(E_2') \ldots P(E_{10}')$ so $= P(E_1 \cup E_2 \cup \ldots \cup E_{10}) = 1 - [.95]^{10} = .401$. Similarly, for 25 points, the desired probability is $= 1 - [P(E')]^{25} = 1 - (.95)^{25} = .723$

77.   Let q denote the probability that a rivet is defective.

    a. P(seam need rework) $= .20 = 1 - $ P(seam doesn't need rework)

               $= 1 - $ P(no rivets are defective)

               $= 1 - $ P(1st isn't def $\cap \ldots \cap 25^{th}$ isn't def)

               $= 1 - (1 - q)^{25}$, so $.80 = (1 - q)^{25}$, $1 - q = (.80)^{1/25}$, and thus $q = 1 - .99111 = .00889$.

    b. The desired condition is $.10 = 1 - (1 - q)^{25}$, i.e. $(1 - q)^{25} = .90$, from which $q = 1 - .99579 = .00421$.

79. Let A_1 = older pump fails, A_2 = newer pump fails, and $x = P(A_1 \cap A_2)$. Then $P(A_1) = .10 + x$, $P(A_2) = .05 + x$, and $x = P(A_1 \cap A_2) = P(A_1) P(A_2) = (.10 + x)(.05 + x)$. The resulting quadratic equation, $x^2 - .85x + .005 = 0$, has roots $x = .0059$ and $x = .8441$. Hopefully the smaller root is the actual probability of system failure.

81.

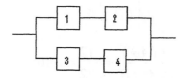

Using the hints, let $P(A_i) = p$, and $x = p^2$, then $P(\text{system lifetime exceeds } t_0) = p^2 + p^2 - p^4 = 2p^2 - p^4 = 2x - x^2$. Now, set this equal to .99, or $2x - x^2 = .99 \Rightarrow x^2 - 2x + .99 = 0$. Solving for x gives $p^2 = x = 0.9$ or 1.1, and so $p = 1.049$ or $.9487$. Since the value we want is a probability, and has to be ≤ 1, we use the value of $p = .9487$.

83. $P(\text{both detect the defect}) = 1 - P(\text{at least one doesn't}) = 1 - .2 = .8$

   a. $P(1^{st} \text{ detects} \cap 2^{nd} \text{ doesn't}) = P(1^{st} \text{ detects}) - P(1^{st} \text{ does} \cap 2^{nd} \text{ does})$
   $$= .9 - .8 = .1$$
   Similarly, $P(1^{st} \text{ doesn't} \cap 2^{nd} \text{ does}) = .1$, so $P(\text{exactly one does}) = .1 + .1 = .2$

   b. $P(\text{neither detects a defect}) = 1 - [P(\text{both do}) + P(\text{exactly 1 does})]$
   $$= 1 - [.8 + .2] = 0$$
   so $P(\text{all 3 escape}) = (0)(0)(0) = 0$.

85.

a. Let D_1 = detection on 1st fixation, D_2 = detection on 2nd fixation.

P(detection in at most 2 fixations) = $P(D_1) + P(D_1' \cap D_2)$

$$= P(D_1) + P(D2 \mid D1')P(D_1)$$
$$= p + p(1-p) = p(2-p).$$

b. Define $D_1, D_2, \ldots, D_n$ as in **a**. Then P(at most n fixations)

$= P(D_1) + P(D_1' \cap D_2) + P(D_1' \cap D_2' \cap D_3) + \ldots + P(D_1' \cap D_2' \cap \ldots \cap D_{n-1}' \cap D_n)$

$= p + p(1-p) + p(1-p)^2 + \ldots + p(1-p)^{n-1}$

$$= p [1 + (1-p) + (1-p)^2 + \ldots + (1-p)^{n-1}] = p \bullet \frac{1-(1-p)^n}{1-(1-p)} = 1-(1-p)^n$$

Alternatively, P(at most n fixations) = $1 - $ P(at least n+1 are req'd)

$$= 1 - \text{P(no detection in } 1^{st} \text{ n fixations)}$$
$$= 1 - P(D_1' \cap D_2' \cap \ldots \cap D_n')$$
$$= 1 - (1-p)^n$$

c. P(no detection in 3 fixations) = $(1-p)^3$

d. P(passes inspection) = P({not flawed} $\cup$ {flawed and passes})

$$= \text{P(not flawed)} + \text{P(flawed and passes)}$$
$$= .9 + \text{P(passes} \mid \text{flawed)} \bullet \text{P(flawed)} = .9 + (1-p)^3(.1)$$

e. P(flawed | passed) = $\dfrac{P(flawed \cap passed)}{P(passed)} = \dfrac{.1(1-p)^3}{.9 + .1(1-p)^3}$

For p = .5, P(flawed | passed) = $\dfrac{.1(.5)^3}{.9 + .1(.5)^3} = .0137$

87. P(system works) = P($1 - 2$ works $\cap$ $3 - 4 - 5 - 6$ works $\cap$ 7 works)

$$= \text{P(} 1 - 2 \text{ works) P(} 3 - 4 - 5 - 6 \text{ works) P(} 7 \text{ works)}$$
$$= (.99) (.9639) (.9) = .8588$$

With the subsystem in figure 2.14 connected in parallel to this subsystem,

P(system works) = $.8588 + .927 - (.8588)(.927) = .9897$

89.

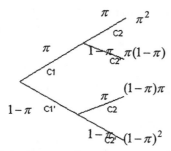

P(at most 1 is lost) = 1 – P(both lost) = $1 - \pi^2$

P(exactly 1 lost) = $2\pi(1 - \pi)$

P(exactly 1 | at most 1) = $\dfrac{P(\text{exactly 1})}{P(\text{at most 1})} = \dfrac{2\pi(1-\pi)}{1-\pi^2} = \dfrac{2\pi}{1+\pi}$

Supplementary Exercises

91.

a. P(line 1) = $\dfrac{500}{1500} = .333$;

P(Crack) = $\dfrac{.50(500)+.44(400)+.40(600)}{1500} = \dfrac{666}{1500} = .444$

b. P(Blemish | line 1) = .15

c. P(Surface Defect) = $\dfrac{.10(500)+.08(400)+.15(600)}{1500} = \dfrac{172}{1500}$

P(line 1 and Surface Defect) = $\dfrac{.10(500)}{1500} = \dfrac{50}{1500}$

So P(line 1 | Surface Defect) = $= \dfrac{50/1500}{172/1500} = .291$

93. $P(A \cup B) = P(A) + P(B) - P(A)P(B) \rightarrow .626 = P(A) + P(B) - .144 \rightarrow P(A) + P(B) = .770$.
So $P(A) + P(B) = .770$ and $P(A)P(B) = .144$.
Let $x = P(A)$ and $y = P(B)$, then using the first equation, $y = .77 - x$, and substituting this
into the second equation, we get $x(.77 - x) = .144$ or $x^2 - .77x + .144 = 0$. Use the

quadratic formula to solve: $\dfrac{.77 \pm \sqrt{.77^2 - (4)(.144)}}{2} = \dfrac{.77 \pm .13}{2} = .32$ or $.45$.

Therefore, $P(A) = .45$ and $P(B) = .32$.

95.

    a.   There are $5 \times 4 \times 3 \times 2 \times 1 = 120$ possible orderings, so $P(BCDEF) = \frac{1}{120} = .0083$

    b.   # orderings in which F is $3^{rd} = 4 \times 3 \times 1 * \times 2 \times 1 = 24$, (* because F must be here), so
        $P(F\ 3^{rd}) = \frac{24}{120} = .2$

    c. $P(F\ last) = \dfrac{4 \times 3 \times 2 \times 1 \times 1}{120} = .2$

97.   When three experiments are performed, there are 3 different ways in which detection can
occur on exactly 2 of the experiments: (i) #1 and #2 and not #3 (ii) #1 and not #2 and
#3; (iii) not#1 and #2 and #3. If the impurity is present, the probability of exactly 2
detections in three (independent) experiments is $(.8)(.8)(.2) + (.8)(.2)(.8) + (.2)(.8)(.8) = .384$. If the impurity is absent, the analogous probability is $3(.1)(.1)(.9) = .027$. Thus

$P(\text{present} \mid \text{detected in exactly 2 out of 3}) = \dfrac{P(\text{detected in exactly } 2 \cap \text{present})}{P(\text{detected in exactly 2})}$

$= \dfrac{(.384)(.4)}{(.384)(.4) + (.027)(.6)} = .905$

99.

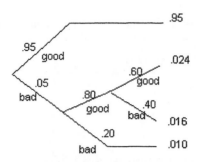

a. P(pass inspection) = P(pass initially $\cup$ passes after recrimping) = P(pass initially) + P(fails initially $\cap$ goes to recrimping $\cap$ is corrected after recrimping)

= .95 + (.05)(.80)(.60) (following path "bad–good–good" on tree diagram)

= .974

b. P(needed no recrimping | passed inspection) = $\dfrac{.95}{.974}$ = .9754

101. Let A = 1st functions, B = 2nd functions, so P(B) = .9, P(A $\cup$ B) = .96, P(A $\cap$ B)=.75. Thus, P(A $\cup$ B) = P(A) + P(B) − P(A $\cap$ B) = P(A) + .9 − .75 = .96, implying P(A) = .81. This gives P(B | A) = $\dfrac{P(B \cap A)}{P(A)} = \dfrac{.75}{.81} = .926$

103.

a. The law of total probability gives

$$P(\text{late}) = \sum_{i=1}^{3} P(\text{late} \mid E_i) \cdot P(E_i)$$

$$= (.02)(.40) + (.01)(.50) + (.05)(.10) = .018$$

b. P(E_1' | on time) = 1 − P(E_1 | on time)

$$= 1 - \frac{P(E_1 \cap \text{on time})}{P(\text{on time})} = 1 - \frac{(.98)(.4)}{.982} = .601$$

105.

    a. P(all different) $= \dfrac{(365)(364)...(356)}{(365)^{10}} = .883 \rightarrow$

    P(at least two the same) $= 1 - .883 = .117$

    b. P(at least two the same) $= .476$ for k=22, and $= .507$ for k=23

    c. P(at least two have the same SS number) $= 1 - $ P(all different) $=$

    $1 - \dfrac{(1000)(999)...(991)}{(1000)^{10}} = 1 - .956 = .044$. Thus P(at least one "coincidence") $=$

    P(BD coincidence $\cup$ SS coincidence) $= .117 + .044 - (.117)(.044) = .156$

107.   P(detection by the end of the nth glimpse) $= 1 - $ P(not detected in 1st n)

    $= 1 - P(G_1' \cap G_2' \cap ... \cap G_n') = 1 - P(G_1')P(G_2') ... P(G_n')$

    $= 1 - (1 - p_1)(1 - p_2) ... (1 - p_n) = 1 - \prod\limits_{i=1}^{n}(1 - p_i)$

109.

    a. P(all in correct room) $= \dfrac{1}{4 \times 3 \times 2 \times 1} = \dfrac{1}{24} = .0417$

    b. The 9 outcomes which yield incorrect assignments are: 2143, 2341, 2413, 3142, 3412, 3421, 4123, 4321, and 4312, so P(all incorrect) $= \dfrac{9}{24} = .375$

111. Note: s = 0 means that the very first candidate interviewed is hired. Each entry below is the candidate hired for the given policy and outcome.

Outcome	s=0	s=1	s=2	s=3	Outcome	s=0	s=1	s=2	s=3
1234	1	4	4	4	3124	3	1	4	4
1243	1	3	3	3	3142	3	1	4	2
1324	1	4	4	4	3214	3	2	1	4
1342	1	2	2	2	3241	3	2	1	1
1423	1	3	3	3	3412	3	1	1	2
1432	1	2	2	2	3421	3	2	2	1
2134	2	1	4	4	4123	4	1	3	3
2143	2	1	3	3	4132	4	1	2	2
2314	2	1	1	4	4213	4	2	1	3
2341	2	1	1	1	4231	4	2	1	1
2413	2	1	1	3	4312	4	3	1	2
2431	2	1	1	1	4321	4	3	2	1

s	0	1	2	3
P(hire#1)	$\frac{6}{24}$	$\frac{11}{24}$	$\frac{10}{24}$	$\frac{6}{24}$

So s = 1 is best.

113. $P(A_1) = P(\text{draw slip 1 or 4}) = \frac{1}{2}$; $P(A_2) = P(\text{draw slip 2 or 4}) = \frac{1}{2}$;
$P(A_3) = P(\text{draw slip 3 or 4}) = \frac{1}{2}$; $P(A_1 \cap A_2) = P(\text{draw slip 4}) = \frac{1}{4}$;
$P(A_2 \cap A_3) = P(\text{draw slip 4}) = \frac{1}{4}$; $P(A_1 \cap A_3) = P(\text{draw slip 4}) = \frac{1}{4}$
Hence $P(A_1 \cap A_2) = P(A_1)P(A_2) = \frac{1}{4}$, $P(A_2 \cap A_3) = P(A_2)P(A_3) = \frac{1}{4}$,
$P(A_1 \cap A_3) = P(A_1)P(A_3) = \frac{1}{4}$, thus there exists pairwise independence
$P(A_1 \cap A_2 \cap A_3) = P(\text{draw slip 4}) = \frac{1}{4} \neq 1/8 = P(A_1)P(A_2)P(A_3)$, so the events are not mutually independent.

CHAPTER 3

Section 3.1

1.

S:	FFF	SFF	FSF	FFS	FSS	SFS	SSF	SSS
X:	0	1	1	1	2	2	2	3

3. M = the difference between the large and the smaller outcome with possible values 0, 1, 2, 3, 4, or 5; W = 1 if the sum of the two resulting numbers is even and W = 0 otherwise, a Bernoulli random variable.

5. No. In the experiment in which a coin is tossed repeatedly until a H results, let Y = 1 if the experiment terminates with at most 5 tosses and Y = 0 otherwise. The sample space is infinite, yet Y has only two possible values.

7.
   a. Possible values are 0, 1, 2, ..., 12; discrete

   b. With N = # on the list, values are 0, 1, 2, ... , N; discrete

   c. Possible values are 1, 2, 3, 4, ... ; discrete

   d. { x: $0 < x < \infty$ } if we assume that a rattlesnake can be arbitrarily short or long; not discrete

   e. With c = amount earned per book sold, possible values are 0, c, 2c, 3c, ... , 10,000c; discrete

   f. { y: $0 < y < 14$} since 0 is the smallest possible pH and 14 is the largest possible pH; not discrete

   g. With m and M denoting the minimum and maximum possible tension, respectively, possible values are { x: $m < x < M$ }; not discrete

   h. Possible values are 3, 6, 9, 12, 15, ... -- i.e. 3(1), 3(2), 3(3), 3(4), ...giving a first element, etc,; discrete

9.

    a. Returns to 0 can occur only after an even number of tosses; possible S values are 2, 4, 6, 8, …(i.e. 2(1), 2(2), 2(3), 2(4),…) an infinite sequence, so x is discrete.

    b. Now a return to 0 is possible after any number of tosses greater than 1, so possible values are 2, 3, 4, 5, … (1+1,1+2, 1+3, 1+4, …, an infinite sequence) and X is discrete

Section 3.2

11.

    a.

x	4	6	8
P(x)	0.45	0.4	0.15

    b.

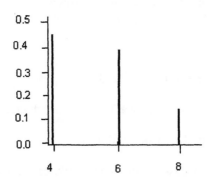

    c. $P(x \geq 6) = .40 + .15 = .55$ $P(x > 6) = .15$

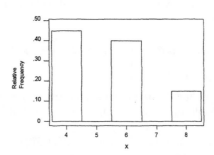

Chapter 3: Discrete Random Variables and Probability Distributions

13.

    a. $P(X \le 3) = p(0) + p(1) + p(2) + p(3) = .10 + .15 + .20 + .25 = .70$

    b. $P(X < 3) = P(X \le 2) = p(0) + p(1) + p(2) = .45$

    c. $P(3 \le X) = p(3) + p(4) + p(5) + p(6) = .55$

    d. $P(2 \le X \le 5) = p(2) + p(3) + p(4) + p(5) = .71$

    e. The number of lines not in use is $6 - X$, so $6 - X = 2$ is equivalent to $X = 4$, $6 - X = 3$ to $X = 3$, and $6 - X = 4$ to $X = 2$. Thus we desire $P(2 \le X \le 4) = p(2) + p(3) + p(4) = .65$

    f. $6 - X \ge 4$ if $6 - 4 \ge X$, i.e. $2 \ge X$, or $X \le 2$, and $P(X \le 2) = .10 + .15 + .20 = .45$

15.

    a. (1,2) (1,3) (1,4) (1,5) (2,3) (2,4) (2,5) (3,4) (3,5) (4,5)

    b. $P(X = 0) = p(0) = P[\{ (3,4) (3,5) (4,5)\}] = \frac{3}{10} = .3$

        $P(X = 2) = p(2) = P[\{ (1,2) \}] = \frac{1}{10} = .1$

        $P(X = 1) = p(1) = 1 - [p(0) + p(2)] = .60$, and $p(x) = 0$ if $x \ne 0, 1, 2$

    c. $F(0) = P(X \le 0) = P(X = 0) = .30$

        $F(1) = P(X \le 1) = P(X = 0 \text{ or } 1) = .90$

        $F(2) = P(X \le 2) = 1$

    The c.d.f. is

$$F(x) = \begin{cases} 0 & x < 0 \\ .30 & 0 \le x < 1 \\ .90 & 1 \le x < 2 \\ 1 & 2 \le x \end{cases}$$

17.

    a. $P(2) = P(Y = 2) = P(1^{st} 2 \text{ batteries are acceptable})$
            $= P(AA) = (.9)(.9) = .81$

    b. $p(3) = P(Y = 3) = P(UAA \text{ or } AUA) = (.1)(.9)^2 + (.1)(.9)^2 = 2[(.1)(.9)^2] = .162$

59

    c. The fifth battery must be an A, and one of the first four must also be an A. Thus,
p(5) = P(AUUUA or UAUUA or UUAUA or UUUAA) = $4[(.1)^3(.9)^2]$ = .00324

    d. P(Y = y) = p(y) = P(the yth is an A and so is exactly one of the first y – 1)
=$(y – 1)(.1)^{y-2}(.9)^2$, y = 2,3,4,5,…

19.   p(0) = P(Y = 0) = P(both arrive on Wed.) = (.3)(.3) = .09
p(1) = P(Y = 1) = P[(W,Th) or (Th,W) or (Th,Th)] = (.3)(.4) + (.4)(.3) + (.4)(.4) = .40
p(2) = P(Y = 2) = P[(W,F) or (Th,F) or (F,W) or (F,Th) or (F,F)] = .32
p(3) = 1 – [.09 + .40 + .32] = .19

21.

    a. Using the formula p(x) = $\log_{10}(1 + 1/x)$ gives the following values: p(1) = .301,
p(2) = .176, p(3) = .125, p(4) = .097, p(5) = .079, p(6) = .067, p(7) = .058, p(8) =
.051, p(9) = .046. The distribution specified by Benford's Law is not uniform on
these nine digits; rather, lower digits (1 and 2) are much more likely to be the
lead digit of a number than higher digits (8 or 9).

    b. The jumps in F(x) occur at 0, … , 8. We display the cumulative probabilities
here: F(1) = .301, F(2) = .477, F(3) = .602, F(4) = .699, F(5) = .778, F(6) = .845,
F(7) = .903, F(8) = .954, F(9) = 1. So, F(x) = 0 for x < 1; F(x) = .301 for 1 ≤ x <
2; F(x) = .477 for 2 ≤ x < 3; etc.

    c. P(X ≤ 3) = F(3) = .602. P(X ≥ 5) = 1 – P(X < 5) = 1 – P(X ≤ 4) = 1 – F(4) = 1 –
.699 = .301.

23.

    a. P(X = 2) = .39 - .19 = .20

    b. P(X > 3) = 1 - .67 = .33

    c. P(2 ≤ X ≤ 5) = .92 - .19 = .78

    d. P(2 < X < 5) = .92 - .39 = .53

25. $P(0) = P(Y = 0) = P(B \text{ first}) = p$
        $P(1) = P(Y = 1) = P(G \text{ first, then B}) = P(GB) = (1 - p)p$
        $P(2) = P(Y = 2) = P(GGB) = (1 - p)^2 p$
        Continuing, $p(y) = P(Y=y) = P(y \text{ G's and then a B}) = (1 - p)^y p$ for $y = 0,1,2,3,\ldots$

27.

a. The sample space consists of all possible permutations of the numbers 1, 2, 3, 4:

outcome	y value	outcome	y value	outcome	y value
1234	4	2314	1	3412	0
1243	2	2341	0	3421	0
1324	2	2413	0	4132	1
1342	1	2431	1	4123	0
1423	1	3124	1	4213	1
1432	2	3142	0	4231	2
2134	2	3214	2	4312	0
2143	0	3241	1	4321	0

b. Thus $p(0) = P(Y = 0) = \frac{9}{24}$, $p(1) = P(Y = 1) = \frac{8}{24}$, $p(2) = P(Y = 2) = \frac{6}{24}$,
$p(3) = P(Y = 3) = 0$, $p(3) = P(Y = 3) = \frac{1}{24}$.

Section 3.3

29.

a. $E(X) = \displaystyle\sum_{x=0}^{4} x \cdot p(x)$
$= (0)(.08) + (1)(.15) + (2)(.45) + (3)(.27) + (4)(.05) = 2.06$

b. $V(X) = \displaystyle\sum_{x=0}^{4} (x - 2.06)^2 \cdot p(x) = (0 - 2.06)^2(.08) + \ldots + (4 - 2.06)^2(.05)$
$= .339488 + .168540 + .001620 + .238572 + .188180 = .9364$

c. $\sigma_x = \sqrt{.9364} = .9677$

d. $V(X) = \left[\displaystyle\sum_{x=0}^{4} x^2 \cdot p(x) \right] - (2.06)^2 = 5.1800 - 4.2436 = .9364$

31. $E(Y) = .60; E(Y^2) = 1.1 \rightarrow V(Y) = E(Y^2) - [E(Y)]^2 = 1.1 - (.60)^2 = .74$

$\sigma_y = \sqrt{.74} = .8602$

$E(Y) \pm \sigma_y = .60 \pm .8602 = (-.2602, 1.4602)$ or $(0, 1)$.

$P(Y = 0) + P(Y = 1) = .85$

33.

a. $E(X^2) = \sum_{x=0}^{1} x^2 \cdot p(x) = (0^2)(1 - p) + (1^2)(p) = (1)(p) = p$

b. $V(X) = E(X^2) - [E(X)]^2 = p - p^2 = p(1 - p)$

c. $E(X^{79}) = (0^{79})(1 - p) + (1^{79})(p) = p$

35. Let $h(X)$ denote the net revenue (sales revenue – order cost) as a function of X. Then $h_3(X)$ and $h_4(X)$ are the net revenue for 3 and 4 copies purchased, respectively. For x = 1 or 2 , $h_3(X) = 2x - 3$, but at x = 3,4,5,6 the revenue plateaus. Following similar reasoning, $h_4(X) = 2x - 4$ for x=1,2,3, but plateaus at 4 for x = 4,5,6.

x	1	2	3	4	5	6
$h_3(x)$	−1	1	3	3	3	3
$h_4(x)$	−2	0	2	4	4	4
p(x)	$\frac{1}{15}$	$\frac{2}{15}$	$\frac{3}{15}$	$\frac{4}{15}$	$\frac{3}{15}$	$\frac{2}{15}$

$E[h_3(X)] = \sum_{x=1}^{6} h_3(x) \cdot p(x) = (-1)(\frac{1}{15}) + \ldots + (3)(\frac{2}{15}) = 2.4667$

Similarly, $E[h_4(X)] = \sum_{x=1}^{6} h_4(x) \cdot p(x) = (-2)(\frac{1}{15}) + \ldots + (4)(\frac{2}{15}) = 2.6667$

Ordering 4 copies gives slightly higher revenue, on the average.

37. $E(X) = \sum_{x=1}^{n} x \cdot \left(\frac{1}{n}\right) = \left(\frac{1}{n}\right) \sum_{x=1}^{n} x = \frac{1}{n}\left[\frac{n(n+1)}{2}\right] = \frac{n+1}{2}$

$E(X^2) = \sum_{x=1}^{n} x^2 \cdot \left(\frac{1}{n}\right) = \left(\frac{1}{n}\right) \sum_{x=1}^{n} x^2 = \frac{1}{n}\left[\frac{n(n+1)(2n+1)}{6}\right] = \frac{(n+1)(2n+1)}{6}$

So $V(X) = \frac{(n+1)(2n+1)}{6} - \left(\frac{n+1}{2}\right)^2 = \frac{n^2-1}{12}$

39. $E(X) = \sum_{x=1}^{4} x \cdot p(x) = 2.3$, $E(X^2) = 6.1$, so $V(X) = 6.1 - (2.3)^2 = .81$

Each lot weighs 5 lbs, so weight left $= 100 - 5X$. Thus the expected weight left is $100 - 5E(X) = 88.5$, and the variance of the weight left is $V(100 - 5X) = V(-5X) = 25V(X) = 20.25$.

41. $V(aX + b) = \sum_x [aX + b - E(aX + b)]^2 \cdot p(x) = \sum_x [aX + b - (a\mu + b)]^2 p(x)$

$= \sum_x [aX - (a\mu)]^2 p(x) = a^2 \sum_x [X - \mu]^2 p(x) = a^2 V(X).$

43. With $a = 1$ and $b = -c$, $E(X - c) = 1E(X) + (-c) = E(X) - c$. When $c = \mu$, $E(X - \mu) = E(X) - \mu = \mu - \mu = 0$, so the expected deviation from the mean is zero.

Section 3.4

45. $a \leq X \leq b$ means that $a \leq x \leq b$ for all x in the domain of X. Hence using the laws of addition, $\sum a \cdot p(x) \leq \sum x \cdot p(x) \leq \sum b \cdot p(x)$, or $a \cdot \sum p(x) \leq \sum x \cdot p(x) \leq b \cdot \sum p(x)$, or $a \leq E(X) \leq b$. This last step uses the facts that $\sum p(x) = 1$, and $\sum x \cdot p(x)$ is the definition of $E(X)$.

47.

   a. $B(4;15,.3) = .515$

   b. $b(4;15,.3) = B(4;15,.3) - B(3;15,.3) = .219$

   c. $b(6;15,.7) = B(6;15,.7) - B(5;15,.7) = .012$

   d. $P(2 \leq X \leq 4) = B(4;15,.3) - B(1;15,.3) = .480$

   e. $P(2 \leq X) = 1 - P(X \leq 1) = 1 - B(1;15,.3) = .965$

   f. $P(X \leq 1) = B(1;15,.7) = .000$

   g. $P(2 < X < 6) = P(2 < X \leq 5) = B(5;15,.3) - B(2;15,.3) = .595$

49. $X \sim \text{Bin}(6, .10)$

    a. $P(X = 1) = \binom{n}{x}(p)^x(1-p)^{n-x} = \binom{6}{1}(.1)^1(.9)^5 = .3543$

    b. $P(X \geq 2) = 1 - [P(X = 0) + P(X = 1)]$.

        From **a**, we know $P(X = 1) = .3543$, and $P(X = 0) = \binom{6}{0}(.1)^0(.9)^6 = .5314$.

        Hence $P(X \geq 2) = 1 - [.3543 + .5314] = .1143$

    c. Either 4 or 5 goblets must be selected

        i)   Select 4 goblets with zero defects: $P(X = 0) = \binom{4}{0}(.1)^0(.9)^4 = .6561$.

        ii) Select 4 goblets, one of which has a defect, and the 5th is good:

$$\left[\binom{4}{1}(.1)^1(.9)^3 \right] \times .9 = .26244$$

        So the desired probability is $.6561 + .26244 = .91854$

51.

    a. $E(X) = np = 25(.25) = 6.25$

    b. $\text{Var}(X) = np(1-p) = 25(.25)(.75) = 4.6875$, so $\text{SD}(X) = 2.165$

    c. $P(X > 6.25 + 2(2.165)) = P(X > 10.58) = 1 - P(X \leq 10.58) = 1 - P(X \leq 10) = 1 - B(10;25,.25) = .030$

53. Let S = has at least one citation. Then $p = .4$, $n = 15$

    a. If at least 10 have no citations (Failure), then at most 5 have had at least one (Success): $P(X \leq 5) = B(5;15,.40) = .403$

    b. $P(X \leq 7) = B(7;15,.40) = .787$

    c. $P(5 \leq X \leq 10) = P(X \leq 10) - P(X \leq 4) = .991 - .217 = .774$

55. Let S represent a telephone that is submitted for service while under warranty and must be replaced. Then $p = P(S) = P(\text{replaced} \mid \text{submitted}) \cdot P(\text{submitted}) = (.40)(.20) = .08$. Thus X, the number among the company's 10 phones that must be replaced, has a binomial distribution with $n = 10$, $p = .08$, so $p(2) = P(X=2) =$

$$\binom{10}{2}(.08)^2(.92)^8 = .1478$$

57. X = the number of flashlights that work. Let B = {battery has acceptable voltage}.
Then P(flashlight works) = P(both batteries work) = P(B)P(B) = (.9)(.9) = .81
We assume that the batteries' voltage levels are independent, so X ~ Bin (10, .81).

$$P(X \geq 9) = P(X=9) + P(X=10) = \binom{10}{9}(.81)^9(.19) + \binom{10}{10}(.81)^{10} = .285 + .122 = .407$$

59.

a. P(rejecting claim when p = .8) = B(15;25,.8) = .017

b. P(not rejecting claim when p = .7) = P(X ≥ 16 when p = .7)
= 1 – B(15;25,.7) = 1 – .189 = .811; for p = .6, this probability is
= 1 – B(15;25,.6) = 1 – .575 = .425.

c. The probability of rejecting the claim when p = .8 becomes B(14;25,.8) = .006,
smaller than in **a** above. However, the probabilities of **b** above increase to .902
and .586, respectively.

61. If topic A is chosen, when n = 2, P(at least half received)
= P(X ≥ 1) = 1 – P(X = 0) = 1 – (.1)² = .99
If B is chosen, when n = 4, P(at least half received)
= P(X ≥ 2) = 1 – P(X ≤ 1) = 1 – (0.1)⁴ – 4(.1)³(.9) = .9963
Thus topic B should be chosen.
If p = .5, the probabilities are .75 for A and .6875 for B, so now A should be chosen.

63.

a. $b(x; n, 1-p) = \binom{n}{x}(1-p)^x(p)^{n-x} = \binom{n}{n-x}(p)^{n-x}(1-p)^x = b(n-x; n, p)$

Alternatively, P(x S's when P(S) = 1 – p) = P(n-x F's when P(F) = p), since the
two events are identical, but the labels S and F are arbitrary and so can be
interchanged (if P(S) and P(F) are also interchanged), yielding P(n-x S's when
P(S) = 1 – p) as desired.

b. B(x;n,1 – p) = P(at most x S's when P(S) = 1 – p)
= P(at least n-x F's when P(F) = p)
= P(at least n-x S's when P(S) = p)
= 1 – P(at most n-x-1 S's when P(S) = p)
= 1 – B(n-x-1;n,p)

c. Whenever p > .5, (1 – p) < .5 so probabilities involving X can be calculated using
the results **a** and **b** in combination with tables giving probabilities only for p ≤ .5

65.

    a. Although there are three payment methods, we are only concerned with S = uses a debit card and F = does not use a debit card. Thus we can use the binomial distribution. So n = 100 and p = .2. $E(X) = np = 100(.5) = 20$, and $V(X) = 16$.

    b. With S = doesn't pay with cash, n = 100 and p = .7, $E(X) = np = 100(.7) = 70$, and $V(X) = 21$.

67. When p = .5, $\mu = 10$ and $\sigma = 2.236$, so $2\sigma = 4.472$ and $3\sigma = 6.708$.
The inequality $|X - 10| \geq 4.472$ is satisfied if either $X \leq 5$ or $X \geq 15$, or $P(|X - \mu| \geq 2\sigma)$ $= P(X \leq 5 \text{ or } X \geq 15) = .021 + .021 = .042$. The inequality $|X - 10| \geq 6.708$ is satisfied if either $X \leq 3$ or $X \geq 17$, so $P(|X - \mu| \geq 3\sigma) = P(X \leq 3 \text{ or } X \geq 17) = .001 + .001 = .002$.

In the case p = .75, $\mu = 15$ and $\sigma = 1.937$, so $2\sigma = 3.874$ and $3\sigma = 5.811$. $P(|X - 15| \geq 3.874) = P(X \leq 11 \text{ or } X \geq 19) = .041 + .024 = .065$, whereas $P(|X - 15| \geq 5.811) = P(X \leq 9) = .004$.

All these probabilities are considerably less than the upper bounds .25 (for k = 2) and .11 (for k = 3) given by Chebyshev.

Section 3.5

69. $X \sim h(x; 6, 12, 7)$

    a. $P(X=5) = \dfrac{\binom{7}{5}\binom{5}{1}}{\binom{12}{6}} = \dfrac{105}{924} = .114$

    b. $P(X \leq 4) = 1 - P(X \geq 5) = 1 - [P(X=5) + P(X=6)] =$

$$1 - \left[\frac{\binom{7}{5}\binom{5}{1}}{\binom{12}{6}} + \frac{\binom{7}{6}}{\binom{12}{6}} \right] = 1 - \frac{105 + 7}{924} = 1 - .121 = .879$$

    c. $E(X) = \left(\dfrac{6 \cdot 7}{12} \right) = 3.5$; $\sigma = \sqrt{\left(\frac{6}{11}\right)\left(6\right)\left(\frac{7}{12}\right)\left(\frac{5}{12}\right)} = \sqrt{.795} = .892$

      $P(X > 3.5 + .892) = P(X > 4.392) = P(X \geq 5) = .121$ (see part **b**)

    d. We can approximate the hypergeometric distribution with the binomial if the population size and the number of successes are large: $h(x;15,40,400)$ approaches $b(x;15,.10)$. So $P(X \le 5) \approx B(5;15,.10)$ from the binomial tables $= .998$

71.

    a. Possible values of X are 5, 6, 7, 8, 9, 10. (In order to have less than 5 of the granite, there would have to be more than 10 of the basaltic).

$$P(X = 5) = h(5; 15,10,20) = \frac{\binom{10}{5}\binom{10}{10}}{\binom{20}{15}} = .0163 \,.$$

Following the same pattern for the other values, we arrive at the pmf, in table form below.

x	5	6	7	8	9	10
p(x)	.0163	.1354	.3483	.3483	.1354	.0163

    b. P(all 10 of one kind or the other) $= P(X = 5) + P(X = 10) = .0163 + .0163 = .0326$

    c. $E(X) = n \cdot \dfrac{M}{N} = 15 \cdot \dfrac{10}{20} = 7.5$; $V(X) = \left(\dfrac{5}{19}\right)(7.5)\left(1 - \dfrac{10}{20}\right) = .9868$; $\sigma_x = .9934$

    $\mu \pm \sigma = 7.5 \pm .9934 = (6.5066, 8.4934)$, so we want
    $P(X = 7) + P(X = 8) = .3483 + .3483 = .6966$

Chapter 3: Discrete Random Variables and Probability Distributions

73.

    a. $h(x; 10,10,20)$ (the successes here are the top 10 pairs, and a sample of 10 pairs is drawn from among the 20)

    b. Let X = the number among the top 5 who play E-W.
Then P(all of top 5 play the same direction) =

$$P(X = 5) + P(X = 0) = h(5;10,5,20) + h(5;10,5,20) = \frac{\binom{15}{5}}{\binom{20}{10}} + \frac{\binom{15}{10}}{\binom{20}{10}} = .033$$

    c. $N = 2n; M = n; n = n$
$h(x;n,n,2n)$

$$E(X) = n \cdot \frac{n}{2n} = \frac{1}{2}n;$$

$$V(X) = \left(\frac{2n-n}{2n-1}\right) \cdot n \cdot \frac{n}{2n} \cdot \left(1 - \frac{n}{2n}\right) = \left(\frac{n}{2n-1}\right) \cdot \frac{n}{2} \cdot \left(1 - \frac{n}{2n}\right) = \left(\frac{n}{2n-1}\right) \cdot \frac{n}{2} \cdot \left(\frac{1}{2}\right)$$

75.

    a. With S = a female child and F = a male child, let X = the number of F's before the 2nd S. Then $P(X = x) = nb(x;2, .5)$

    b. P(exactly 4 children) = P(exactly 2 males)
$$= nb(2;2,.5) = (3)(.0625) = .188$$

    c. P(at most 4 children) = $P(X \le 2)$

$$= \sum_{x=0}^{2} nb(x;2,.5) = .25 + 2(.25)(.5) + 3(.0625) = .688$$

    d. $E(X) = \frac{(2)(.5)}{.5} = 2$, so the expected # of children = $E(X + 2) = E(X) + 2 = 4$

77. This is identical to an experiment in which a single family has children until exactly 6 females have been born(since p = .5 for each of the three families), so p(x) = $nb(x;6,.5)$ and $E(X) = 6$ (= 2+2+2, the sum of the expected number of males born to each one.)

Section 3.6

79.

    a. $P(X \le 8) = F(8;5) = .932$

    b. $P(X = 8) = F(8;5) - F(7;5) = .065$

    c. $P(X \ge 9) = 1 - P(X \le 8) = .068$

    d. $P(5 \le X \le 8) = F(8;5) - F(4;5) = .492$

    e. $P(5 < X < 8) = F(7;5) - F(5;5) = .867 - .616 = .251$

81.

    a. $P(X \le 10) = F(10;20) = .011$

    b. $P(X > 20) = 1 - F(20;20) = 1 - .559 = .441$

    c. $P(10 \le X \le 20) = F(20;20) - F(9;20) = .559 - .005 = .554$
       $P(10 < X < 20) = F(19;20) - F(10;20) = .470 - .011 = .459$

    d. $E(X) = \lambda = 20$, $\sigma_X = \sqrt{\lambda} = 4.472$, so $P(\mu - 2\sigma < X < \mu + 2\sigma)$
       $= P(20 - 8.944 < X < 20 + 8.944) = P(11.056 < X < 28.944)$
       $= P(X \le 28) - P(X \le 11) = F(28;20) - F(11;20) = .966 - .021 = .945$

83. $p = \dfrac{1}{200}$; $n = 1000$; $\lambda = np = 5$

    a. $P(5 \le X \le 8) = F(8;5) - F(4;5) = .492$

    b. $P(X \ge 8) = 1 - P(X \le 7) = 1 - .867 = .133$

85.

    a. $\lambda = 8$ when $t = 1$, so $P(X = 6) = F(6;8) - F(5;8) = .313 - .191 = .122$,
       $P(X \ge 6) = 1 - F(5;8) = .809$, and $P(X \ge 10) = 1 - F(9;8) = .283$

    b. $t = 90$ min $= 1.5$ hours, so $\lambda = 12$; thus the expected number of arrivals is 12 and
       the SD $= \sqrt{12} = 3.464$

    c. $t = 2.5$ hours implies that $\lambda = 20$; in this case, $P(X \ge 20) = 1 - F(19;20) = .530$
       and $P(X \le 10) = F(10;20) = .011$.

87.

    a. For a two hour period the parameter of the distribution is $\lambda = \alpha t = (4)(2) = 8$, so $P(X = 10) = F(10;8) - F(9;8) = .099$.

    b. For a 30 minute period, $\alpha t = (4)(.5) = 2$, so $P(X = 0) = F(0;2) = .135$

    c. $E(X) = \alpha t = 2$

89. $\alpha = 1/(\text{mean time between occurrences}) = 2$

    a. $\alpha t = (2)(2) = 4$

    b. $P(X > 5) \; 1 - P(X \le 5) = 1 - .785 = .215$

    c. Solve for t, given $\alpha = 2$: $0.1 = e^{-2t} \rightarrow t = 0.5\ln(0.1) \approx 1.15$ years

91.

    a. For a one-quarter acre plot, the parameter is $(80)(.25) = 20$, so $P(X \le 16) = F(16;20) = .221$

    b. The expected number of trees is $\alpha \cdot (\text{area}) = 80(85,000) = 6,800,000$.

    c. The area of the circle is $\pi r^2 = .031416$ sq. miles or 20.106 acres. Thus X has a Poisson distribution with parameter $\lambda = \alpha(20.106) = 1608.5$.

93.

    a. No events in $(0, t+\Delta t)$ if and only if no events in $(0, t)$ and no events in $(t, t+\Delta t)$. Thus, $P_0(t+\Delta t) = P_0(t) \cdot P(\text{no events in } (t, t+\Delta t)) = P_0(t)[1 - \alpha \cdot \Delta t - o(\Delta t)]$

    b. $$\frac{P_0(t + \Delta t) - P_0(t)}{\Delta t} = -\alpha P_0(t) - P_0(t) \cdot \frac{o(\Delta t)}{\Delta t}$$

    c. $\frac{d}{dt}\left[e^{-\alpha t}\right] = -\alpha e^{-\alpha t} = -\alpha P_0(t)$, as desired.

    d. $$\frac{d}{dt}\left[\frac{e^{-\alpha t}(\alpha t)^k}{k!}\right] = \frac{-\alpha e^{-\alpha t}(\alpha t)^k}{k!} + \frac{k\alpha e^{-\alpha t}(\alpha t)^{k-1}}{k!} =$$

    $-\alpha \frac{e^{-\alpha t}(\alpha t)^k}{k!} + \alpha \frac{e^{-\alpha t}(\alpha t)^{k-1}}{(k-1)!} = -\alpha P_k(t) + \alpha P_{k-1}(t)$ as desired.

Supplementary Exercises

95.

a. p(1) = P(exactly one suit) = P(all spades) + P(all hearts) + P(all diamonds)

$$+ \text{ P(all clubs)} = 4\text{P(all spades)} = 4 \cdot \frac{\binom{13}{5}}{\binom{52}{5}} = .00198$$

p(2) = P(all hearts and spades with at least one of each) + ...+ P(all diamonds and clubs with at least one of each)
= 6 P(all hearts and spades with at least one of each)
= 6 [P(1 h and 4 s) + P(2 h and 3 s) + P(3 h and 2 s) + P(4 h and 1 s)]

$$= 6 \cdot \left[2 \cdot \frac{\binom{13}{4}\binom{13}{1}}{\binom{52}{5}} + 2 \cdot \frac{\binom{13}{3}\binom{13}{2}}{\binom{52}{5}} \right] = 6 \left[\frac{18,590 + 44,616}{2,598,960} \right] = .14592$$

$$p(4) = 4\text{P(2 spades, 1 h, 1 d, 1 c)} = \frac{4 \cdot \binom{13}{2}(13)(13)(13)}{\binom{52}{5}} = .26375$$

p(3) = 1 − [p(1) + p(2) + p(4)] = .58835

b. $\mu = \sum_{x=1}^{4} x \cdot p(x) = 3.114, \ \sigma^2 = \left[\sum_{x=1}^{4} x^2 \cdot p(x) \right] - (3.114)^2 = .405, \sigma = .636$

97.

a. b(x;15,.75)

b. P(X > 10) = 1 - B(10;15, .75) = 1 - .314 = .686

c. B(10;15, .75) - B(5;15, .75) = .314 - .001 = .313

d. $\mu = (15)(.75) = 11.75, \sigma^2 = (15)(.75)(.25) = 2.81$

e. Requests can all be met if and only if X ≤ 10, and 15 − X ≤ 8, i.e. if 7 ≤ X ≤ 10, so P(all requests met) = B(10; 15,.75) - B(6; 15,.75) = .310

99.   Let $X \sim \text{Bin}(5, .9)$. Then $P(X \geq 3) = 1 - P(X \leq 2) = 1 - B(2;5,.9) = .991$

101.

    a. $N = 500$, $p = .005$, so $np = 2.5$ and $b(x; 500, .005) \approx p(x; 2.5)$, a Poisson p.m.f.

    b. $P(X = 5) = p(5; 2.5) - p(4; 2.5) = .9580 - .8912 = .0668$

    c. $P(X \geq 5) = 1 - p(4;2.5) = 1 - .8912 = .1088$

103.   Let Y denote the number of tests carried out.
For $n = 3$, possible Y values are 1 and 4.
$P(Y = 1) = P(\text{no one has the disease}) = (.9)^3 = .729$ and $P(Y = 4) = .271$,
so $E(Y) = (1)(.729) + (4)(.271) = 1.813$, as contrasted with the 3 tests necessary
without group testing.
For $n = 5$, possible values of Y are 1 and 6.
$P(Y = 1) = (.9)^5 = .5905$, so $P(Y = 6) = .4095$ and $E(Y) = (1)(.5905) + (6)(.4095) = 3.0475$, less than the 5 tests necessary without group testing.

105. $p(2) = P(X = 2) = P(\text{S on \#1 and S on \#2}) = p^2$
$p(3) = P(\text{S on \#3 and S on \#2 and F on \#1}) = (1-p)p^2$
$p(4) = P(\text{S on \#4 and S on \#3 and F on \#2}) = (1-p)p^2$
$p(5) = P(\text{S on \#5 and S on \#4 and F on \#3 and no 2 consecutive S's on trials prior to \#3}) = [1 - p(2)](1-p)p^2$
$p(6) = P(\text{S on \#6 and S on \#5 and F on \#4 and no 2 consecutive S's on trials prior to \#4}) = [1 - p(2) - p(3)](1-p)p^2$
In general, for $x = 5, 6, 7, \ldots$: $p(x) = [1 - p(2) - \ldots - p(x-3)](1-p)p^2$
For $p = .9$,

x	2	3	4	5	6	7	8
p(x)	.81	.081	.081	.0154	.0088	.0023	.0010

So $P(X \leq 8) = p(2) + \ldots + p(8) = .9995$

107.

    a. Let event C = seed carries single spikelets, and event P = seed produces ears with single spikelets. Then $P(P \cap C) = P(P \mid C) \cdot P(C) = .29 \,(.40) = .116$. Let X = the number of seeds out of the 10 selected that meet the condition $P \cap C$. Then X ~ Bin(10, .116). $P(X = 5) = \binom{10}{5}(.116)^5 (.884)^5 = .002857$

    b. For 1 seed, the event of interest is P = seed produces ears with single spikelets. $P(P) = P(P \cap C) + P(P \cap C') = .116 + P(P \mid C') \cdot P(C') = .116 + (.26)(.60) = .272$. Let Y = the number out of the 10 seeds that meet condition P. Then Y ~ Bin(10, .272), and $P(Y = 5) = .0767$. $P(Y \le 5) = b(0;10,.272) + \ldots + b(5;10,.272) = .041813 + \ldots + .076719 = .97024$

109.

    a. $P(X = 0) = F(0;2) = 0.135$

    b. Let S = an operator who receives no requests. Then p = .135 and we wish P(4 S's in 5 trials) $= b(4;5,.135) = \binom{5}{4}(.135)^4 (.865)^1 = .00144$

    c. P(all receive x) = P(first receives x) $\cdot \ldots \cdot$ P(fifth receives x) $= \left[\dfrac{e^{-2} 2^x}{x!} \right]^5$, and

    P(all receive the same number) is the sum from x = 0 to ∞.

111.   The number sold is min (X, 5), so $E[\min(x, 5)] = \sum_{}^{\infty} \min(x,5) p(x;4)$

$$= (0)p(0;4) + (1)\, p(1;4) + (2)\, p(2;4) + (3)\, p(3;4) + (4)\, p(4;4) + 5\sum_{x=5}^{\infty} p(x;4)$$

$$= 1.735 + 5[1 - F(4;4)] = 3.59$$

113.

    a. No; probability of success is not the same for all tests

    b. There are four ways exactly three could have positive results. Let D represent those with the disease and D' represent those without the disease.

Combination		**Probability**
D	D'	
0	3	$$\left[\binom{5}{0}(.2)^0(.8)^5\right]\cdot\left[\binom{5}{3}(.9)^3(.1)^2\right]$$ $= (.32768)(.0729) = .02389$
1	2	$$\left[\binom{5}{1}(.2)^1(.8)^4\right]\cdot\left[\binom{5}{2}(.9)2(.1)^3\right]$$ $= (.4096)(.0081) = .00332$
2	1	$$\left[\binom{5}{2}(.2)^2(.8)^3\right]\cdot\left[\binom{5}{1}(.9)^1(.1)^4\right]$$ $= (.2048)(.00045) = .00009216$
3	0	$$\left[\binom{5}{3}(.2)^3(.8)^2\right]\cdot\left[\binom{5}{0}(.9)^0(.1)^5\right]$$ $= (.0512)(.00001) = .000000512$

Adding up the probabilities associated with the four combinations yields 0.0273.

115.

a. $p(x;\lambda,\mu) = \frac{1}{2}p(x;\lambda) + \frac{1}{2}p(x;\mu)$ where both $p(x;\lambda)$ and $p(x;\mu)$ are Poisson p.m.f.'s and thus ≥ 0, so $p(x;\lambda,\mu) \geq 0$. Further,

$$\sum_{x=0}^{\infty} p(x;\lambda,\mu) = \frac{1}{2}\sum_{x=0}^{\infty} p(x;\lambda) + \frac{1}{2}\sum_{x=0}^{\infty} p(x;\mu) = \frac{1}{2} + \frac{1}{2} = 1$$

b. $.6p(x;\lambda) + .4p(x;\mu)$

c. $E(X) = \sum_{x=0}^{\infty} x[\frac{1}{2}p(x;\lambda) + \frac{1}{2}p(x;\mu)] = \frac{1}{2}\sum_{x=0}^{\infty} xp(x;\lambda) + \frac{1}{2}\sum_{x=0}^{\infty} xp(x;\mu)$

$= \frac{1}{2}\lambda + \frac{1}{2}\mu = \frac{\lambda+\mu}{2}$

d. $E(X^2) = \frac{1}{2}\sum_{x=0}^{\infty} x^2 p(x;\lambda) + \frac{1}{2}\sum_{x=0}^{\infty} x^2 p(x;\mu) = \frac{1}{2}(\lambda^2 + \lambda) + \frac{1}{2}(\mu^2 + \mu)$ since for a

Poisson r.v., $E(X^2) = V(X) + [E(X)]^2 = \lambda + \lambda^2 \rightarrow$

$V(X) = \frac{1}{2}[\lambda^2 + \lambda + \mu^2 + \mu] - \left[\frac{\lambda+\mu}{2}\right]^2 = \left(\frac{\lambda-\mu}{2}\right)^2 + \frac{\lambda+\mu}{2}$

117. $P(X = j) = \sum_{i=1}^{10} P\,(\text{arm on track } i \cap X = j) = \sum_{i=1}^{10} P\,(X = j \mid \text{arm on } i) \cdot p_i$

$= \sum_{i=1}^{10} P\,(\text{next seek at i+j+1 or i-j-1}) \cdot p_i = \sum_{i=1}^{10}(p_{i+j+1} + p_{i-j-1})p_i$

where $p_k = 0$ if $k < 0$ or $k > 10$.

119. Let $A = \{x: |x - \mu| \geq k\sigma\}$.

Then $\sigma^2 = \sum_{all\,x}(x - \mu)^2 p(x) \geq \sum_{A}(x - \mu)^2 p(x) \geq (k\sigma)^2 \sum_{A} p(x)$.

But $\sum_{A} p(x) = P(X \text{ is in } A) = P(|X-\mu| \geq k\sigma)$, so $\sigma^2 \geq k^2\sigma^2 \cdot P(|X-\mu| \geq k\sigma)$,

so $P(|X-\mu| \geq k\sigma) \leq 1/k^2$.

121.

a. Let A_1 = {voice}, A_2 = {data}, and X = duration of a call.
Then $E(X) = E(X|A_1)P(A_1) + E(X|A_2)P(A_2) = 3(.75) + 1(.25) = 2.5$ minutes

b. Let X = the number of chips in a cookie.
Then $E(X) = E(X|i=1)P(i=1) + E(X|i=2)P(i=2) + E(X|i=3)P(i=3)$.
If X is Poisson, then its mean is the specified λ – that is, $E(X|i) = i+1$.
Therefore, $E(X) = 2(.20) + 3(.50) + 4(.30) = 3.1$ chips

CHAPTER 4

Section 4.1

1.

    a. $P(X \le 1) = \int_{-\infty}^{1} f(x)dx = \int_{0}^{1} \frac{1}{2}x\,dx = \frac{1}{4}x^2 \Big]_{0}^{1} = .25$

    b. $P(.5 \le X \le 1.5) = \int_{.5}^{1.5} \frac{1}{2}x\,dx = \frac{1}{4}x^2 \Big]_{.5}^{1.5} = .5$

    c. $P(X > 1.5) = \int_{.5}^{\infty} f(x)dx = \int_{.5}^{2} \frac{1}{2}x\,dx = \frac{1}{4}x^2 \Big]_{1.5}^{2} = \frac{7}{16} \approx .438$

3.

    a.

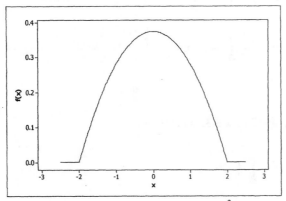

    b. $P(X > 0) = \int_{0}^{2} .09375(4 - x^2)dx = .09375(4x - \frac{x^3}{3}) \Big]_{0}^{2} = .5$

    c. $P(-1 < X < 1) = \int_{-1}^{1} .09375(4 - x^2)dx = .6875$

    d. $P(X < -.5 \text{ or } X > .5) = 1 - P(-.5 \le X \le .5) = 1 - \int_{-.5}^{.5} .09375(4 - x^2)dx = 1 - .3672 =$

       $.6328$

5.

    a. $1 = \int_{-\infty}^{\infty} f(x)dx = \int_{0}^{2} kx^2 dx = k\left(\frac{x^3}{3}\right)\Big|_{0}^{2} = k\left(\frac{8}{3}\right) \Rightarrow k = \frac{3}{8}$

    b. $P(0 \le X \le 1) = \int_{0}^{1} \frac{3}{8}x^2 dx = \frac{1}{8}x^3\Big|_{0}^{1} = \frac{1}{8} = .125$

    c. $P(1 \le X \le 1.5) = \int_{1}^{1.5} \frac{3}{8}x^2 dx = \frac{1}{8}x^3\Big|_{1}^{1.5} = \frac{1}{8}\left(\frac{3}{2}\right)^3 - \frac{1}{8}(1)^3 = \frac{19}{64} \approx .2969$

    d. $P(X \ge 1.5) = 1 - \int_{0}^{1.5} \frac{3}{8}x^2 dx = \frac{1}{8}x^3\Big|_{0}^{1.5} = 1 - \left[\frac{1}{8}\left(\frac{3}{2}\right)^3 - 0\right] = 1 - \frac{27}{64} = \frac{37}{64} \approx .5781$

7.

    a. $f(x) = \frac{1}{10}$ for $25 \le x \le 35$ and $= 0$ otherwise

    b. $P(X > 33) = \int_{33}^{35} \frac{1}{10} dx = .2$

    c. $E(X) = \int_{25}^{35} x \cdot \frac{1}{10} dx = \frac{x^2}{20}\Big|_{25}^{35} = 30$

    30 ± 2 is from 28 to 32 minutes:

    $P(28 < X < 32) = \int_{28}^{32} \frac{1}{10} dx = \frac{1}{10}x\Big|_{28}^{32} = .4$

    d. $P(a \le x \le a+2) = \int_{a}^{a+2} \frac{1}{10} dx = .2$, since the interval has length 2.

78

9.

    a. $P(X \le 6) = = \int_5^6 .15e^{-.15(x-5)} dx = .15 \int_0^{.5} e^{-.15u} du$ (after $u = x - .5$) =

    $e^{-.15u} \Big|_0^{.5} = 1 - e^{-.825} \approx .562$

    b. $1 - .562 = .438; .438$

    c. $P(5 \le Y \le 6) = P(Y \le 6) - P(Y \le 5) \approx .562 - .491 = .071$

Section 4.2

11.

    a. $P(X \le 1) = F(1) = \frac{1}{4} = .25$

    b. $P(.5 \le X \le 1) = F(1) - F(.5) = \frac{3}{16} = .1875$

    c. $P(X > .5) = 1 - P(X \le .5) = 1 - F(.5) = \frac{15}{16} = .9375$

    d. $.5 = F(\tilde{\mu}) = \frac{\tilde{\mu}^2}{4} \Rightarrow \tilde{\mu}^2 = 2 \Rightarrow \tilde{\mu} = \sqrt{2} \approx 1.414$

    e. $f(x) = F'(x) = \frac{x}{2}$ for $0 \le x < 2$, and $= 0$ otherwise

    f. $E(X) = \int_{-\infty}^{\infty} x \cdot f(x)dx = \int_0^2 x \cdot \frac{1}{2} x dx = \frac{1}{2} \int_0^2 x^2 dx = \frac{x^3}{6} \Big|_0^2 = \frac{8}{6} \approx 1.333$

    g. $E(X^2) = \int_{-\infty}^{\infty} x^2 f(x)dx = \int_0^2 x^2 \frac{1}{2} x dx = \frac{1}{2} \int_0^2 x^3 dx = \frac{x^4}{8} \Big|_0^2 = 2,$

    So $Var(X) = E(X^2) - [E(X)]^2 = 2 - \left(\frac{8}{6}\right)^2 = \frac{8}{36} \approx .222, \sigma_x \approx .471$

    h. From **g**, $E(X^2) = 2$

13.

a. $1 = \int_0^\infty \frac{k}{x^4} dx = \frac{k}{3} \Rightarrow k = 3$

b. $F(x) = \int_{-\infty}^x f(y) dy = \int_1^x 3y^{-4} dy = -x^{-3} + 1 = 1 - \frac{1}{x^3}$. So $F(x) = \begin{cases} 0, & x \le 1 \\ 1 - x^{-3}, & x > 1 \end{cases}$

c. $P(x > 2) = 1 - F(2) = 1 - \left(1 - \frac{1}{8}\right) = \frac{1}{8}$ or .125;
$P(2 < x < 3) = F(3) - F(2) = \left(1 - \frac{1}{27}\right) - \left(1 - \frac{1}{8}\right) = .963 - .875 = .088$

d. $E(X) = \int_0^\infty x\left(\frac{3}{x^4}\right) dx = \int_0^\infty \left(\frac{3}{x^3}\right) dx = \frac{3}{2}$; $E(X^2) = \int_0^\infty x^2\left(\frac{3}{x^4}\right) dx = \int_0^\infty \left(\frac{3}{x^2}\right) dx = 3$

$V(X) = E(X^2) - [E(X)]^2 = 3 - \left(\frac{3}{2}\right)^2 = 3 - \frac{9}{4} = \frac{3}{4}$; $\sigma = \sqrt{V(x)} = \sqrt{\frac{3}{4}} = .866$

e. $P(1.5 - .866 < x < 1.5 + .866) = P(x < 2.366) = F(2.366) = 1 - (2.366^{-3}) = .9245$

15.

a. The pdf of X appears below.

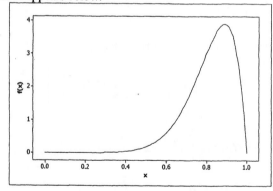

80

As for the cdf, $F(x) = 0$ for $x \le 0$, $= 1$ for $x \ge 1$, and for $0 < x < 1$,

$$F(x) = \int_{-\infty}^{x} f(y)dy = \int_{0}^{x} 90y^8(1-y)dy = 90 \int_{0}^{x} (y^8 - y^9)dy =$$

$$90\left(\tfrac{1}{9}y^9 - \tfrac{1}{10}y^{10}\right)\Big|_{0}^{x} = 10x^9 - 9x^{10}$$

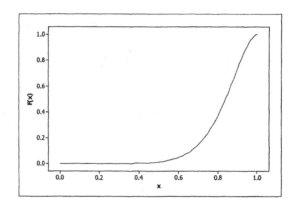

b. $F(.5) = 10(.5)^9 - 9(.5)^{10} \approx .0107$

c. $P(.25 \le X \le .5) = F(.5) - F(.25) \approx .0107 - [10(.25)^9 - 9(.25)^{10}]$
$$\approx .0107 - .0000 \approx .0107$$

d. The 75th percentile is the value of x for which $F(x) = .75 \Rightarrow .75 = 10x^9 - 9x^{10} \Rightarrow x \approx .9036$

e. $E(X) = \int_{-\infty}^{\infty} x \cdot f(x)dx = \int_{0}^{1} x \cdot 90x^8(1-x)dx = 90 \int_{0}^{1} x^9(1-x)dx$

$$= 9x^{10} - \tfrac{90}{11}x^{11}\Big|_{0}^{1} = \tfrac{9}{11} \approx .8182$$

$$E(X^2) = \int_{-\infty}^{\infty} x^2 \cdot f(x)dx = \int_{0}^{1} x^2 \cdot 90x^8(1-x)dx = 90 \int_{0}^{1} x^{10}(1-x)dx$$

$$= \tfrac{90}{11}x^{11} - \tfrac{90}{12}x^{12}\Big|_{0}^{1} \approx .6818$$

Therefore, $V(X) \approx .6818 - (.8182)^2 = .0124$, and $\sigma_x = .11134$

f. $\mu \pm \sigma = (.7068, .9295)$. Thus, $P(\mu - \sigma \le X \le \mu + \sigma) = F(.9295) - F(.7068) = .8465 - .1602 = .6863$, so the probability X is *more* than 1 sd from its mean equals $1 - .6863 = .3137$.

17.

    a. $F(X) = \dfrac{x - A}{B - A} = p \Rightarrow x = (100p)\text{th percentile} = A + (B - A)p$

    b. $E(X) = \displaystyle\int_A^B x \cdot \dfrac{1}{B - A} dx = \dfrac{1}{B - A} \cdot \dfrac{x^2}{2} \Big]_A^B = \dfrac{1}{2} \cdot \dfrac{1}{B - A} \cdot \left(B^2 - A^2\right) = \dfrac{A + B}{2}$

         $E(X^2) = \dfrac{1}{3} \cdot \dfrac{1}{B - A} \cdot \left(B^3 - A^3\right) = \dfrac{A^2 + AB + B^2}{3}$

         $V(X) = \left(\dfrac{A^2 + AB + B^2}{3}\right) - \left(\dfrac{(A + B)}{2}\right)^2 = \dfrac{(B - A)^2}{12}, \quad \sigma_x = \dfrac{(B - A)}{\sqrt{12}}$

    c. $E(X^n) = \displaystyle\int_A^B x^n \cdot \dfrac{1}{B - A} dx = \dfrac{B^{n+1} - A^{n+1}}{(n+1)(B - A)}$

19.

    a. $P(X \le 1) = F(1) = .25[1 + \ln(4)] \approx .597$

    b. $P(1 \le X \le 3) = F(3) - F(1) \approx .966 - .597 \approx .369$

    c. $f(x) = F'(x) = .25 \ln(4) - .25 \ln(x)$ for $0 < x < 4$

21. $E(\text{area}) = E(\pi R^2) = \displaystyle\int_{-\infty}^{\infty} \pi r^2 f(r) dr = \int_9^{11} \pi r^2 \left(\dfrac{3}{4}\right)\left(1 - (10 - r)^2\right) dr = \dfrac{501}{5}\pi = 314.79$

23. With X = temperature in °C, temperature in °F $= \dfrac{9}{5}X + 32,$ so

    $E\left[\dfrac{9}{5}X + 32\right] = \dfrac{9}{5}(120) + 32 = 248; \quad Var\left[\dfrac{9}{5}X + 32\right] = \left(\dfrac{9}{5}\right)^2 \cdot (2)^2 = 12.96$, so $\sigma = 3.6$

25.

    a. $P(Y \le 1.8\tilde{\mu} + 32) = P(1.8X + 32 \le 1.8\tilde{\mu} + 32) = P(X \le \tilde{\mu}) = .5$

    b. 90th for $Y = 1.8\eta(.9) + 32$ where $\eta(.9)$ is the 90th percentile for X, since
    $P(Y \le 1.8\eta(.9) + 32) = P(1.8X + 32 \le 1.8\eta(.9) + 32) = (X \le \eta(.9)) = .9$ as desired.

    c. The (100p)th percentile for Y is $1.8\eta(p) + 32$, verified by substituting p for .9 in the argument of **b**. When $Y = aX + b$, (i.e. a linear transformation of X), and the (100p)th percentile of the X distribution is $\eta(p)$, then the corresponding (100p)th percentile of the Y distribution is $a \cdot \eta(p) + b$. (same linear transformation applied to X's percentile)

27.   Since X is uniform on [0,360], $E(X) = \dfrac{0 + 360}{2} = 180$ and $\sigma_X = \dfrac{360 - 0}{\sqrt{12}} = 30\sqrt{12}$. Using the linear representation of Y, $E(Y) = (2\pi/360)E(X) - \pi = (2\pi/360)(180) - \pi = 0$, and $\sigma_Y = (2\pi/360)\sigma_X = (2\pi/360)(30\sqrt{12}) = \dfrac{\pi\sqrt{12}}{6} \approx 1.814$. (In fact, Y is uniform on $[-\pi, \pi]$.)

Section 4.3

29.

    a. .9838 is found in the 2.1 row and the .04 column of the z table, so c = 2.14

    b. $P(0 \le Z \le c) = .291 \Rightarrow \Phi(c) = .7910 \Rightarrow c = .81$

    c. $P(c \le Z) = .121 \Rightarrow 1 - P(c \le Z) = P(Z < c) = \Phi(c) = 1 - .121 = .8790 \Rightarrow c = 1.17$

    d. $P(-c \le Z \le c) = \Phi(c) - \Phi(-c) = \Phi(c) - (1 - \Phi(c)) = 2\Phi(c) - 1 \Rightarrow \Phi(c) = .9920$
        $\Rightarrow c = .97$

    e. $P(c \le |Z|) = .016 \Rightarrow 1 - .016 = .9840 = 1 - P(c \le |Z|) = P(|Z| < c) = P(-c < Z < c) = \Phi(c) - \Phi(-c) = 2\Phi(c) - 1 \Rightarrow \Phi(c) = .9920 \Rightarrow c = 2.41$

31.

    a. Area under Z curve above $z_{.0055}$ is .0055, which implies that $\Phi(z_{.0055}) = 1 - .0055 = .9945$, so $z_{.0055} = 2.54$

    b. $\Phi(z_{.09}) = .9100 \Rightarrow z = 1.34$ (since .9099 appears as the 1.34 entry).

    c. $\Phi(z_{.663}) = $ area below $z_{.663} = .3370 \Rightarrow z_{.633} \approx -.42$

33.

    a. $P(X \leq 18) = P\left(z \leq \dfrac{18-15}{1.25} \right) = P(Z \leq 2.4) = \Phi(2.4) = .9452$

    b. $P(10 \leq X \leq 12) = P(-4.00 \leq Z \leq -2.40) \approx P(Z \leq -2.40) = \Phi(-2.40) = .0082$

    c. $P(|X - 15| \leq 1.5(1.25)) = P(|Z| \leq 1.5) = P(-1.5 \leq Z \leq 1.5) = 2\Phi(1.5) - 1 = .8664$

35.

    a. $P(X \geq 10) = P(Z \geq .43) = 1 - \Phi(.43) = 1 - .6664 = .3336.$
       $P(X > 10) = P(X \geq 10) = .3336$, since for any continuous distribution, $P(x = a) = 0$.

    b. $P(X > 20) = P(Z > 4) \approx 0$

    c. $P(5 \leq X \leq 10) = P(-1.36 \leq Z \leq .43) = \Phi(.43) - \Phi(-1.36) = .6664 - .0869 = .5795$

    d. $P(8.8 - c \leq X \leq 8.8 + c) = .98$, so $8.8 - c$ and $8.8 + c$ are at the 1st and the 99th percentile of the given distribution, respectively. The 1st percentile of the standard normal distribution has the value -2.33, so $8.8 - c = \mu + (-2.33)\sigma = 8.8 - 2.33(2.8)$ $\Rightarrow c = 2.33(2.8) = 6.524.$

    e. From a, $P(x > 10) = .3336$. Define event A as {diameter > 10}, then P(at least one $A_i) = 1 - P(\text{no } A_i) = 1 - P(A')^4 = 1 - (1 - .3336)^4 = 1 - .1972 = .8028$

37.

    a. $P(X = 105) = 0$, since the normal distribution is continuous;
       $P(X < 105) = P(Z < 0.2) = P(Z \leq 0.2) = \Phi(0.2) = .5793;$
       $P(X \leq 105) = .5793$ as well, since X is continuous

    b. No, the answer does not depend on μ or σ. For any normal rv, $P(|X-\mu| > \sigma) =$
       $P(|Z| > 1) = P(Z < -1 \text{ or } Z > 1) = 2\Phi(-1) = 2(.1587) = .3174$

    c. From the table, $\Phi(z) = .1\% = .001 \rightarrow z \approx -3.09 \rightarrow x = 104 - 3.09(5) = 88.55$. The smallest .1% of chloride concentration values are those less than 88.55 mmol/L

39.

    a. $\mu + \sigma \cdot (91^{st}$ percentile from std normal$) = 30 + 5(1.34) = 36.7$

    b. $30 + 5(-1.555) = 22.225$

    c. $\mu = 3.000$ μm; $\sigma = 0.140$. We desire the 90th percentile: $30 + 1.28(0.14) = 3.179$

Chapter 4: Continuous Random Variables and Probability Distributions

41. $P(\text{damage}) = P(X < 100) = P\left(z < \dfrac{100 - 200}{300}\right) = P(Z < -3.33) = .0004.$

$P(\geq 1 \text{ among five is damaged}) = 1 - P(\text{none damaged}) = 1 - (.9996)^5 = 1 - .998 = .002$

43. Since 1.28 is the 90th z percentile ($z_{.1} = 1.28$) and -1.645 is the 5th z percentile ($z_{.05} = 1.645$), the given information implies that $\mu + \sigma(1.28) = 10.256$ and $\mu + \sigma(-1.645) = 9.671$, from which $\sigma(-2.925) = -.585$, $\sigma = .2000$, and $\mu = 10$.

45. With $\mu = .500$ inches, the acceptable range for the diameter is between .496 and .504 inches, so unacceptable bearings will have diameters smaller than .496 or larger than .504. The new distribution has $\mu = .499$ and $\sigma = .002$. $P(X < .496 \text{ or } X > .504) =$

$P\left(Z < \dfrac{.496 - .499}{.002}\right) + P\left(Z > \dfrac{.504 - .499}{.002}\right) = P(Z < -1.5) + P(Z > 2.5) = \Phi(-1.5) + [1 - \Phi(2.5)] = .073$. 7.3% of the bearings will be unacceptable.

47. The stated condition implies that 99% of the area under the normal curve with $\mu = 12$ and $\sigma = 3.5$ is to the left of $c - 1$, so $c - 1$ is the 99th percentile of the distribution. Thus $c - 1 = \mu + \sigma(2.33) = 20.155$, and $c = 21.155$.

49. $X \sim N(3432, 482)$

a. $P(x > 4000) = P\left(Z > \dfrac{4000 - 3432}{482}\right) = P(z > 1.18) = 1 - \Phi(1.18) = 1 - .8810 = .1190$;

$P(3000 < x < 4000) = P\left(\dfrac{3000 - 3432}{482} < Z < \dfrac{4000 - 3432}{482}\right)$

$= \Phi(1.18) - \Phi(-.90) = .8810 - .1841 = .6969$

b. $P(X < 2000 \cup X > 5000) = P\left(Z < \dfrac{2000 - 3432}{482}\right) + P\left(Z > \dfrac{5000 - 3432}{482}\right)$

$= \Phi(-2.97) + [1 - \Phi(3.25)] = .0015 + .0006 = .0021$

c. We will use the conversion 1 lb = 454 g, then 7 lbs = 3178 grams, and we wish to find $P(x > 3178) = P\left(Z > \dfrac{3178 - 3432}{482}\right) = 1 - \Phi(-.53) = .7019$

    d. We need the top .0005 and the bottom .0005 of the distribution. Using the Z table, both .9995 and .0005 have multiple z values, so we will use a middle value, ± 3.295. Then $3432 \pm (482)3.295 = 1844$ and 5020, or the most extreme .1% of all birth weights are less than 1844 g and more than 5020 g.

    e. Converting to lbs yields mean 7.5595 and sd 1.0608. Then

$$P(X > 7) = P\left(Z > \frac{7 - 7.5595}{1.0608}\right) = 1 - \Phi(-.53) = .7019 \quad \text{This yields the same answer as}$$

in part **c.**

51. $P(|X - \mu| \geq \sigma) = P(X \leq \mu - \sigma \text{ or } X \geq \mu + \sigma) = 1 - P(\mu - \sigma \leq X \leq \mu + \sigma)$
$= 1 - P(-1 \leq Z \leq 1) = .3174.$
Similarly, $P(|X - \mu| \geq 2\sigma) = 1 - P(-2 \leq Z \leq 2) = .0456$ and $P(|X - \mu| \geq 3\sigma) = .0026.$
These are considerably less than the bounds 1, .25, and .11 given by Chebyshev.

53. $p = .5 \rightarrow \mu = 12.5 \ \& \ \sigma^2 = 6.25; \ p = .6 \rightarrow \mu = 15 \ \& \ \sigma^2 = 6; \ p = .8 \rightarrow \mu = 20$ and $\sigma^2 = 4$

    a.

p	$P(15 \leq X \leq 20)$	$P(14.5 \leq \text{Normal} \leq 20.5)$
.5	= .212	= $P(.80 \leq Z \leq 3.20)$ = .2112
.6	= .577	= $P(-.20 \leq Z \leq 2.24)$ = .5668
.8	= .573	= $P(-2.75 \leq Z \leq .25)$ = .5957

    b.

p	$P(X \leq 15)$	$P(\text{Normal} \leq 15.5)$
.5	= .885	= $P(Z \leq 1.20)$ = .8849
.6	= .575	= $P(Z \leq .20)$ = .5793
.8	= .017	= $P(Z \leq -2.25)$ = .0122

    c.

p	$P(X \geq 20)$	$P(\text{Normal} \geq 19.5)$
.5	= .002	= $P(Z \geq 2.80)$ = .0026
.6	= .029	= $P(Z \geq 1.84)$ = .0329
.8	= .617	= $P(Z \geq -0.25)$ = .5987

55. $n = 500$, $p = .75 \rightarrow \mu = 375$, $\sigma = 9.68246$

  a. $P(360 \leq X \leq 400) = P(359.5 \leq X \leq 400.5) = P(-1.60 \leq Z \leq 2.58) = .9409$

  b. $P(X < 400) = P(X \leq 399.5) = P(Z \leq 2.53) = .9943$

57.

  a. $F_y(y) = P(Y \leq y) = P(aX + b \leq y) = P\left(X \leq \dfrac{(y-b)}{a} \right)$ for $a > 0$.

  Now differentiate with respect to y to obtain

  $f_y(y) = F_y^{'}(y) = \dfrac{1}{\sqrt{2\pi}\,a\sigma}\, e^{-\frac{1}{2a^2\sigma^2}[y-(a\mu+b)]^2}$ so Y is normal with mean $a\mu + b$

  and variance $a^2\sigma^2$.

  b. Normal, mean $\frac{9}{5}(115) + 32 = 239$, variance $= 12.96$

Section 4.4

59.

  a. $E(X) = \dfrac{1}{\lambda} = 1$

  b. $\sigma = \dfrac{1}{\lambda} = 1$

  c. $P(X \leq 4) = 1 - e^{-(1)(4)} = 1 - e^{-4} = .982$

  d. $P(2 \leq X \leq 5) = 1 - e^{-(1)(5)} - \left[1 - e^{-(1)(2)} \right] = e^{-2} - e^{-5} = .129$

61. mean $= \dfrac{1}{\lambda} = 25{,}000$ implies $\lambda = .00004$

  a. $P(X > 20{,}000) = 1 - P(X \leq 20{,}000) = 1 - F(20{,}000; .00004) = e^{-(.00004)(20{,}000)} = .449$

  $P(X \leq 30{,}000) = F(30{,}000; .00004) = 1 - e^{-1.2} = .699$

  $P(20{,}000 \leq X \leq 30{,}000) = .699 - .551 = .148$

  b. $\sigma = \dfrac{1}{\lambda} = 25{,}000$, so $P(X > \mu + 2\sigma) = P(x > 75{,}000) = 1 - F(75{,}000; .00004) = .05$.

  Similarly, $P(X > \mu + 3\sigma) = P(x > 100{,}000) = .018$

63.

   a. If a customer's calls are typically short, the first calling plan makes more sense. If a customer's calls are somewhat longer, then the second plan makes more sense, viz. 99¢ is less than 20min(10¢/min) = $2 for the first 20 minutes under the first (flat–rate) plan.

   b. $h_1(X) = 10X$, while $h_2(X) = 99$ for $X \le 20$ and $99 + 10(X - 20)$ for $X > 20$. With $\mu = 1/\lambda$ for the exponential distribution, it's obvious that $E[h_1(X)] = 10E[X] = 10\mu$. On the other hand,

$$E[h_2(X)] = 99 + 10\int_{20}^{\infty}(x-20)\lambda e^{-\lambda x}\,dx = 99 + \frac{10}{\lambda}e^{-20\lambda} = 99 + 10\mu e^{-20/\mu}.$$

When $\mu = 10$, $E[h_1(X)] = 100¢ = \$1.00$ while $E[h_2(X)] = 99 + 100e^{-2} \approx \$1.13$.
When $\mu = 15$, $E[h_1(X)] = 150¢ = \$1.50$ while $E[h_2(X)] = 99 + 150e^{-4/3} \approx \$1.39$.
As predicted, the first plan is better when expected call length is lower, and the second plan is better when expected call length is somewhat higher.

65.

   a. $P(X \le 5) = F(5;7) = .238$

   b. $P(X < 5) = P(X \le 5) = .238$

   c. $P(X > 8) = 1 - P(X \le 8) = 1 - F(8;7) = .313$

   d. $P(3 \le X \le 8) = F(8;7) - F(3;7) = .653$

   e. $P(3 < X < 8) = .653$

   f. $P(X < 4 \text{ or } X > 6) = 1 - P(4 \le X \le 6) = 1 - [F(6;7) - F(4;7)] = .713$

67.

   a. $P(12 \le X \le 24) = F(4;4) - F(2;4) = .424$

   b. $P(X \le 24) = F(4;4) = .567$, so while the mean is 24, the median is less than 24, since $P(X \le \tilde{\mu}) = .5$. This is a result of the positive skew of the gamma distribution.

   c. We want a value for which $F(x;4) = .99$. In table A.4, we see $F(10;4) = .990$. So with $\beta = 6$, the 99th percentile $= 6(10) = 60$.

   d. We want a value for which $F(t;4) = .995$. In the table, $F(11;4) = .995$, so $t = 6(11) = 66$. At 66 weeks, only .5% of all transistors would still be operating.

69.

    a. $\{X \geq t\} = A_1 \cap A_2 \cap A_3 \cap A_4 \cap A_5$

    b. $P(X \geq t) = P(A_1) \cdot P(A_2) \cdot P(A_3) \cdot P(A_4) \cdot P(A_5) = \left(e^{-\lambda t}\right)^5 = e^{-.05t} \rightarrow$

        $F_x(t) = P(X \leq t) = 1 - e^{-.05t}$, $f_x(t) = .05e^{-.05t}$ for $t \geq 0$. Thus X also has an
        exponential distribution, but with parameter $\lambda = .05$.

    c. By the same reasoning, $P(X \leq t) = 1 - e^{-n\lambda t}$, so X has an exponential distribution
        with parameter $n\lambda$.

71.

    a. $\{X^2 \leq y\} = \left\{-\sqrt{y} \leq X \leq \sqrt{y}\right\}$

    b. $F_Y(y) = P(X^2 \leq y) = \int_{-\sqrt{y}}^{\sqrt{y}} \frac{1}{\sqrt{2\pi}} e^{-z^2/2} dz$, so

        $f_Y(y) = \frac{1}{\sqrt{2\pi}} e^{-y/2} \frac{1}{2} y^{-1/2} - \frac{1}{\sqrt{2\pi}} e^{-y/2} \frac{-1}{2} y^{-1/2} = \frac{1}{\sqrt{2\pi}} e^{-y/2} y^{-1/2}$.

        We recognize this as the chi–squared pdf with $v = 1$.

Section 4.5

73.

    a. $P(X \le 250) = F(250; 2.5, 200) = 1 - e^{-(250/200)^{2.5}} = 1 - e^{-1.75} \approx .8257$

        $P(X < 250) = P(X \le 250) \approx .8257$

        $P(X > 300) = 1 - F(300; 2.5, 200) = e^{-(1.5)^{2.5}} = .0636$

    b. $P(100 \le X \le 250) = F(250; 2.5, 200) - F(100; 2.5, 200) \approx .8257 - .162 = .6637$

    c. The median $\widetilde{\mu}$ is requested. The equation $F(\widetilde{\mu}) = .5$ reduces to

$$.5 = e^{-(\widetilde{\mu}/200)^{2.5}}, \text{ i.e., } \ln(.5) \approx -\left(\frac{\widetilde{\mu}}{200}\right)^{2.5}, \text{ so } \widetilde{\mu} = (.6931)^4(200) = 172.727.$$

75.   Using the substitution y $= \left(\dfrac{x}{\beta}\right)^{\alpha}$ and dy $= \dfrac{\alpha x^{\alpha-1}}{\beta^{\alpha}} dx$, $\mu = \displaystyle\int_0^{\infty} x \cdot \dfrac{\alpha}{\beta^{\alpha}} x^{\alpha-1} e^{-(x/\beta)^{\alpha}} dx =$

$\beta \displaystyle\int_0^{\infty} y^{1/\alpha} e^{-y} dy = \beta \cdot \Gamma\left(1 + \dfrac{1}{\alpha}\right)$ by definition of the gamma function.

77.

    a. $E(X) = e^{\left(\mu + \frac{\sigma^2}{2}\right)} = e^{4.82} = 123.97$

        $V(X) = \left(e^{(2(4.5)+.8^2)}\right) \cdot \left(e^{.8} - 1\right) = (15,367.34)(.8964) = 13,776.53 \quad \sigma = 117.373$

    b. $P(X \le 100) = P\left(Z \le \dfrac{\ln(100) - 4.5}{.8}\right) = \Phi(0.13) = .5517$

    c. $P(X \ge 200) = P\left(Z \ge \dfrac{\ln(200) - 4.5}{.8}\right) = 1 - \Phi(1.00) = 1 - .8413 = .1587 = P(X > 200)$

79.

a. $E(X) = e^{3.5+(1.2)^2/2} = 68.0335$; $V(X) = e^{2(3.5)+(1.2)^2} \cdot \left(e^{(1.2)^2} - 1\right) = 14907.168 \rightarrow$

$\sigma_x = 122.0949$

b. $P(50 \le X \le 250) = P\left(Z \le \dfrac{\ln(250)-3.5}{1.2}\right) - P\left(Z \le \dfrac{\ln(50)-3.5}{1.2}\right) =$

$P(Z \le 1.68) - P(Z \le .34) = .9535 - .6331 = .3204$

c. $P(X \le 68.0335) = P\left(Z \le \dfrac{\ln(68.0335)-3.5}{1.2}\right) = P(Z \le .60) = .7257$. The lognormal

distribution is not a symmetric distribution.

81.

a. $E(X) = e^{5+(.01)/2} = e^{5.005} = 149.157$; $Var(X) = e^{10+(.01)} \cdot \left(e^{.01} - 1\right) = 223.594$

b. $P(X > 125) = 1 - P(X \le 125) = 1 - P\left(Z \le \dfrac{\ln(125)-5}{.1}\right) = 1 - \Phi(-1.72) = .9573$

c. $P(110 \le X \le 125) = \Phi(-1.72) - \Phi\left(\dfrac{\ln(110)-5}{.1}\right) = .0427 - .0013 = .0414$

d. $\tilde{\mu} = e^5 = 148.41$

e. P(any particular one has $X > 125$) = .9573 $\Rightarrow$ expected # = $10(.9573) = 9.573$

f. We wish the 5th percentile, which is $e^{5+(-1.645)(.1)} = 125.90$

83. The point of symmetry must be $\frac{1}{2}$, so we require that $f\left(\frac{1}{2} - \mu\right) = f\left(\frac{1}{2} + \mu\right)$, i.e.,

$\left(\frac{1}{2} - \mu\right)^{\alpha-1}\left(\frac{1}{2} + \mu\right)^{\beta-1} = \left(\frac{1}{2} + \mu\right)^{\alpha-1}\left(\frac{1}{2} - \mu\right)^{\beta-1}$, which in turn implies that $\alpha = \beta$.

85.

a. $E(X) = \int_0^1 x \cdot \dfrac{\Gamma(\alpha+\beta)}{\Gamma(\alpha)\Gamma(\beta)} x^{\alpha-1}(1-x)^{\beta-1}\,dx = \dfrac{\Gamma(\alpha+\beta)}{\Gamma(\alpha)\Gamma(\beta)} \int_0^1 x^{\alpha}(1-x)^{\beta-1}\,dx =$

$\dfrac{\Gamma(\alpha+\beta)}{\Gamma(\alpha)\Gamma(\beta)} \cdot \dfrac{\Gamma(\alpha+1)\Gamma(\beta)}{\Gamma(\alpha+\beta+1)} = \dfrac{\alpha\Gamma(\alpha)}{\Gamma(\alpha)\Gamma(\beta)} \cdot \dfrac{\Gamma(\alpha+\beta)}{(\alpha+\beta)\Gamma(\alpha+\beta)} = \dfrac{\alpha}{\alpha+\beta}$

b. $E[(1-X)^m] = \int_0^1 (1-x)^m \cdot \dfrac{\Gamma(\alpha+\beta)}{\Gamma(\alpha)\Gamma(\beta)} x^{\alpha-1}(1-x)^{\beta-1}\,dx = \dfrac{\Gamma(\alpha+\beta)}{\Gamma(\alpha)\Gamma(\beta)} \int_0^1 x^{\alpha-1}(1-x)^{m+\beta-1}\,dx =$

$\dfrac{\Gamma(\alpha+\beta)\cdot\Gamma(m+\beta)}{\Gamma(\alpha+\beta+m)\Gamma(\beta)}$. For $m = 1$, $E(1-X) = \dfrac{\beta}{\alpha+\beta}$.

Section 4.6

87. The given probability plot is quite linear, and thus it is quite plausible that the tension distribution is normal.

89. The z percentile values are as follows: $-1.86, -1.32, -1.01, -0.78, -0.58, -0.40, -0.24,$ $-0.08, 0.08, 0.24, 0.40, 0.58, 0.78, 1.01, 1.30,$ and 1.86. The accompanying probability plot is reasonably straight, and thus it would be reasonable to use estimating methods that assume a normal population distribution.

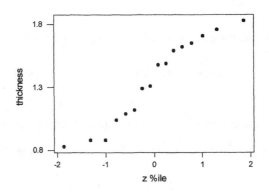

91. The (z percentile, observation) pairs are (–1.66, .736), (–1.32, .863), (–1.01, .865), (–.78, .913), (–.58, .915), (–.40, .937), (–.24, .983), (–.08, 1.007), (.08, 1.011), (.24, 1.064), (.40, 1.109), (.58, 1.132), (.78, 1.140), (1.01, 1.153), (1.32, 1.253), (1.86, 1.394). The accompanying probability plot is straight, suggesting that an assumption of population normality is plausible.

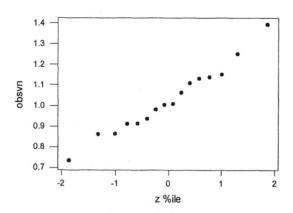

93. To check for plausibility of a lognormal population distribution for the rainfall data of Exercise 81 in Chapter 1, take the natural logs and construct a normal probability plot. This plot and a normal probability plot for the original data appear below. Clearly the log transformation gives quite a straight plot, so lognormality is plausible. The curvature in the plot for the original data implies a positively skewed population distribution – like the lognormal distribution.

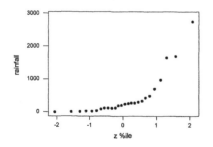 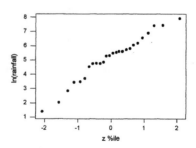

95. The pattern in the plot (below, generated by Minitab) is reasonably linear. By visual inspection alone, it is plausible that strength is normally distributed.

Normal Probability Plot

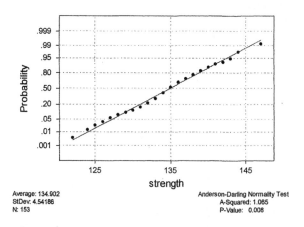

Average: 134.902
StDev: 4.54186
N: 153

Anderson-Darling Normality Test
A-Squared: 1.065
P-Value: 0.008

97. The $(100p)^{th}$ percentile $\eta(p)$ for the exponential distribution with $\lambda = 1$ satisfies $F(\eta(p)) = 1 - \exp[-\eta(p)] = p$, i.e., $\eta(p) = -\ln(1 - p)$. With $n = 16$, we need $\eta(p)$ for $p = \frac{5}{16}, \frac{1.5}{16}, ..., \frac{15.5}{16}$. These are .032, .398, .170, .247, .330, .421, .521, .633, .758, .901, 1.068, 1.269, 1.520, 1.856, 2.367, 3.466. This plot exhibits substantial curvature, casting doubt on the assumption of an exponential population distribution. Because λ is a scale parameter (as is σ for the normal family), $\lambda = 1$ can be used to assess the plausibility of the entire exponential family.

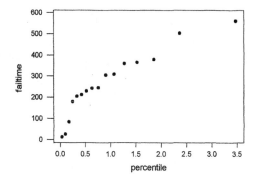

94

Supplementary Exercises

99.

a. For $0 \le Y \le 25$, $F(y) = \dfrac{1}{24} \int_0^y \left(u - \dfrac{u^2}{12} \right) = \dfrac{1}{24} \left(\dfrac{u^2}{2} - \dfrac{u^3}{36} \right) \Big]_0^y$. Thus

$$F(y) = \begin{cases} 0 & y < 0 \\ \dfrac{1}{48}\left(y^2 - \dfrac{y^3}{18} \right) & 0 \le y \le 12 \\ 1 & y > 12 \end{cases}$$

b. $P(Y \le 4) = F(4) = .259$, $P(Y > 6) = 1 - F(6) = .5$
   $P(4 \le X \le 6) = F(6) - F(4) = .5 - .259 = .241$

c. $E(Y) = \dfrac{1}{24} \int_0^{12} y^2 \left(1 - \dfrac{y}{12} \right) dy = \dfrac{1}{24} \left[\dfrac{y^3}{3} - \dfrac{y^4}{48} \right]_0^{12} = 6$

   $E(Y^2) = \dfrac{1}{24} \int_0^{12} y^3 \left(1 - \dfrac{y}{12} \right) dy = 43.2$, so $V(Y) = 43.2 - 36 = 7.2$

d. $P(Y < 4 \text{ or } Y > 8) = 1 - P(4 \le X \le 8) = .518$

e. the shorter segment has length $\min(Y, 12 - Y)$ so

   $E[\min(Y, 12 - Y)] = \int_0^{12} \min(y,12 - y) \cdot f(y)dy = \int_0^{6} \min(y,12 - y) \cdot f(y)dy$

   $+ \int_6^{12} \min(y,12 - y) \cdot f(y)dy = \int_0^{6} y \cdot f(y)dy + \int_6^{12} (12 - y) \cdot f(y)dy = \dfrac{90}{24} = 3.75$

101.

a.   By differentiation, $f(x) = \begin{cases} x^2 & 0 \le x < 1 \\ \dfrac{7}{4} - \dfrac{3}{4}x & 1 \le x \le \dfrac{7}{3} \\ 0 & otherwise \end{cases}$

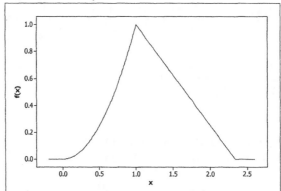

b. $P(.5 \le X \le 2) = F(2) - F(.5) = 1 - \dfrac{1}{2}\left(\dfrac{7}{3} - 2\right)\left(\dfrac{7}{4} - \dfrac{3}{4} \cdot 2\right) - \dfrac{(.5)^3}{3} = \dfrac{11}{12} = .917$

c. $E(X) = \int_0^1 x \cdot x^2 dx + \int_1^{7/3} x \cdot \left(\dfrac{7}{4} - \dfrac{3}{4}x\right) dx = \dfrac{131}{108} = 1.213$

103. $\mu = 137.2$ oz.; $\sigma = 1.6$ oz

a. $P(X > 135) = 1 - \Phi\left(\dfrac{135 - 137.2}{1.6}\right) = 1 - \Phi(-1.38) = 1 - .0838 = .9162$

b.   With Y = the number among ten that contain more than 135 oz, Y ~ Bin(10, .9162).
So, $P(Y \ge 8) = b(8; 10, .9162) + b(9; 10, .9162) + b(10; 10, .9162) = .9549$

c. $\mu = 137.2$; $\dfrac{135 - 137.2}{\sigma} = -1.65 \Rightarrow \sigma = 1.33$

105.

   a. $P(X > 100) = 1 - \Phi\left(\dfrac{100-96}{14}\right) = 1 - \Phi(.29) = 1 - .6141 = .3859$

   b. $P(50 < X < 80) = \Phi\left(\dfrac{80-96}{14}\right) - \Phi\left(\dfrac{50-96}{14}\right)$
   $= \Phi(-1.5) - \Phi(-3.29) = .1271 - .0005 = .1266.$

   c. $a = 5^{th}$ percentile $= 96 + (-1.645)(14) = 72.97.$ $b = 95^{th}$ percentile $= 96 + (1.645)(14)$
   $= 119.03.$ The interval (72.97, 119.03) contains the central 90% of all grain sizes.

107.

   a.

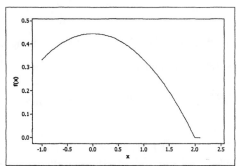

   b. $F(x) = 0$ for $x < -1$ or $= 1$ for $x > 2$. For $-1 \le x \le 2$,

   $$F(x) = \int_{-1}^{x} \frac{1}{9}\left(4 - y^2\right)dy = \frac{1}{9}\left(4x - \frac{x^3}{3}\right) + \frac{11}{27}$$

   c. The median is 0 iff $F(0) = .5$. Since $F(0) = \frac{11}{27}$, this is not the case. Because $\frac{11}{27} < .5$, the median must be greater than 0.

   d. Y is a binomial r.v. with $n = 10$ and $p = P(X > 1) = 1 - F(1) = \frac{5}{27}$

109.

a. $P(X \le 150) = \exp\left[-\exp\left(\dfrac{-(150-150)}{90}\right)\right] = \exp[-\exp(0)] = \exp(-1) = .368$.

$P(X \le 300) = \exp[-\exp(-1.6667)] = .828$, and $P(150 \le X \le 300) = .828 - .368 = .460$.

b. The desired value c is the 90th percentile, so c satisfies

$.9 = \exp\left[-\exp\left(\dfrac{-(c-150)}{90}\right)\right]$. Taking the natural log of each side twice in succession

yields $\ln[-\ln(.9)] = \dfrac{-(c-150)}{90}$, so c = 90(2.250367) + 150 = 352.53.

c. $f(x) = F'(X) = \dfrac{1}{\beta} \cdot \exp\left[-\exp\left(\dfrac{-(x-\alpha)}{\beta}\right)\right] \cdot \exp\left(\dfrac{-(x-\alpha)}{\beta}\right)$

d. We wish the value of x for which f(x) is a maximum; this is the same as the value of

x for which $\ln[f(x)]$ is a maximum. The equation of $\dfrac{d[\ln(f(x))]}{dx} = 0$ gives

$\exp\left(\dfrac{-(x-\alpha)}{\beta}\right) = 1$, so $\dfrac{-(x-\alpha)}{\beta} = 0$, which implies that x = α. Thus the mode is α.

e. E(X) = .5772β + α = 201.95, whereas the mode is 150 and the median is
−(90)ln[−ln(.5)] + 150 = 182.99. The distribution is positively skewed.

111.

a. From a graph of f(x; μ, σ) or by differentiation, x* = μ.

b. No; the density function has constant height for A ≤ X ≤ B.

c. F(x;λ) is largest for x = 0 (the derivative at 0 does not exist since f is not continuous
there) so x* = 0.

d. $\ln[f(x;\alpha,\beta)] = -\ln(\beta^{\alpha}) - \ln(\Gamma(\alpha)) + (\alpha-1)\ln(x) - \dfrac{x}{\beta}; \dfrac{d}{dx}\ln[f(x;\alpha,\beta)] = \dfrac{\alpha-1}{x} - \dfrac{1}{\beta} \Rightarrow$

$x = x^* = (\alpha-1)\beta$

e. From **d** $x^* = \left(\dfrac{\nu}{2}-1\right)(2) = \nu - 2$.

113.

a. Clearly $f(x; \lambda_1, \lambda_2, p) \geq 0$ for all x, and $\int_{-\infty}^{\infty} f(x; \lambda_1, \lambda_2, p)dx$

$$= \int_0^{\infty} \left[p\lambda_1 e^{-\lambda_1 x} + (1-p)\lambda_2 e^{-\lambda_2 x} \right]dx = p\int_0^{\infty} \lambda_1 e^{-\lambda_1 x}dx + (1-p)\int_0^{\infty} \lambda_2 e^{-\lambda_2 x}dx$$

$$= p + (1-p) = 1$$

b. For $x > 0$, $F(x; \lambda_1, \lambda_2, p) = \int_0^x f(y; \lambda_1, \lambda_2, p)dy = p(1 - e^{-\lambda_1 x}) + (1-p)(1 - e^{-\lambda_2 x})$.

c. $E(X) = \int_0^{\infty} x \cdot \left[p\lambda_1 e^{-\lambda_1 x} + (1-p)\lambda_2 e^{-\lambda_2 x} \right]dx$

$$= p\int_0^{\infty} x\lambda_1 e^{-\lambda_1 x}dx + (1-p)\int_0^{\infty} x\lambda_2 e^{-\lambda_2 x}dx = \frac{p}{\lambda_1} + \frac{(1-p)}{\lambda_2}$$

d. $E(X^2) = \frac{2p}{\lambda_1^2} + \frac{2(1-p)}{\lambda_2^2}$, so $Var(X) = \frac{2p}{\lambda_1^2} + \frac{2(1-p)}{\lambda_2^2} - \left[\frac{p}{\lambda_1} + \frac{(1-p)}{\lambda_2} \right]^2$

e. For an exponential r.v., $CV = \dfrac{1/\lambda}{1/\lambda} = 1$. For X hyperexponential,

$$CV = \left[\frac{\frac{2p}{\lambda_1^2} + \frac{2(1-p)}{\lambda_2^2}}{\left[\frac{p}{\lambda_1} + \frac{(1-p)}{\lambda_2} \right]^2} - 1 \right]^{1/2} = \left[\frac{2\left(p\lambda_2^2 + (1-p)\lambda_1^2 \right)}{\left(p\lambda_2 + (1-p)\lambda_1 \right)^2} - 1 \right]^{1/2} = [2r - 1]^{1/2}, \text{ where}$$

$$r = \frac{\left(p\lambda_2^2 + (1-p)\lambda_1^2 \right)}{\left(p\lambda_2 + (1-p)\lambda_1 \right)^2}.$$

But straightforward algebra shows that $r > 1$ provided $\lambda_1 \neq \lambda_2$, so that $CV > 1$.

f. $\mu = \dfrac{n}{\lambda}$, $\sigma^2 = \dfrac{n}{\lambda^2}$, so $\sigma = \dfrac{\sqrt{n}}{\lambda}$ and $CV = \dfrac{1}{\sqrt{n}} < 1$ if $n > 1$.

115.

a. A lognormal distribution, since $\ln\left(\dfrac{I_o}{I_i}\right)$ is a normal r.v.

b. $P\left(I_o > 2I_i\right) = P\left(\dfrac{I_o}{I_i} > 2\right) = P\left(\ln\left(\dfrac{I_o}{I_i}\right) > \ln 2\right) = 1 - P\left(\ln\left(\dfrac{I_o}{I_i}\right) \le \ln 2\right)$

$1 - \Phi\left(\dfrac{\ln 2 - 1}{.05}\right) = 1 - \Phi(-6.14) = 1$

c. $E\left(\dfrac{I_o}{I_i}\right) = e^{1 + .0025/2} = 2.72, \quad Var\left(\dfrac{I_o}{I_i}\right) = e^{2 + .0025} \cdot \left(e^{.0025} - 1\right) = .0185$

117. $F(y) = P(Y \le y) = P(\sigma Z + \mu \le y) = P\left(Z \le \dfrac{(y - \mu)}{\sigma}\right) = \int_{-\infty}^{\frac{(y-\mu)}{\sigma}} \dfrac{1}{\sqrt{2\pi}} e^{-\frac{1}{2}z^2} \, dz$. Now

differentiate with respect to y to obtain a normal pdf with parameters μ and σ.

119.

a. $Y = -\ln(X) \Rightarrow x = e^{-y} = k(y)$, so $k'(y) = -e^{-y}$. Thus since $f(x) = 1$,
$g(y) = 1 \cdot |-e^{-y}| = e^{-y}$ for $0 < y < \infty$, so y has an exponential distribution with parameter $\lambda = 1$.

b. $y = \sigma Z + \mu \Rightarrow y = h(z) = \sigma Z + \mu \Rightarrow z = k(y) = \dfrac{(y - \mu)}{\sigma}$ and $k'(y) = \dfrac{1}{\sigma}$, from which the result follows easily.

c. $y = h(x) = cx \Rightarrow x = k(y) = \dfrac{y}{c}$ and $k'(y) = \dfrac{1}{c}$, from which the result follows easily.

121.

a. Assuming independence, P(all 3 births occur on March 11) = $\left(\frac{1}{365}\right)^3$ = .00000002

b. $\left(\frac{1}{365}\right)^3 (365)$ = .0000073

c. Let X = deviation from due date. X~N(0, 19.88). Then the baby due on March 15 was 4 days early. P(x = –4) ≈ P(–4.5 < x < –3.5)

$= \Phi\left(\frac{-3.5}{19.88}\right) - \Phi\left(\frac{-4.5}{19.88}\right) = \Phi(-.18) - \Phi(-.237) = .4286 - .4090 = .0196$. Similarly, the baby due on April 1 was 21 days early, and P(x = –21)

$\approx \Phi\left(\frac{-20.5}{19.88}\right) - \Phi\left(\frac{-21.5}{19.88}\right) = \Phi(-1.03) - \Phi(-1.08) = .1515 - .1401 = .0114$.

The baby due on April 4 was 24 days early, and P(x = –24) ≈ .0097

Again, assuming independence, P(all 3 births occurred on March 11) = $(.0196)(.0114)(.0097) = .00002145$

d. To calculate the probability of the three births happening on any day, we could make similar calculations as in part c for each possible day, and then add the probabilities.

123.

a. $F_X(x) = P\left(-\frac{1}{\lambda}\ln(1-U) \le x\right) = P(\ln(1-U) \ge -\lambda x) = P\left(1 - U \ge e^{-\lambda x}\right)$

$= P\left(U \le 1 - e^{-\lambda x}\right) = 1 - e^{-\lambda x}$ since $F_U(u)$ = u (U is uniform on [0, 1]). Thus X has an exponential distribution with parameter λ.

b. By taking successive random numbers u_1, u_2, u_3, …and computing $x_i = -\frac{1}{10}\ln(1 - u_i)$, we obtain a sequence of values generated from an exponential distribution with parameter λ = 10.

125. g(μ) + g'(μ)(X–μ) ≤ g(X) implies that E[g(μ) + g'(μ)(X–μ)] = E(g(μ)) + g'(μ)E[(X–μ)] = E(g(μ)) + g'(μ)·0 = g(μ) ≤ E(g(X)), i.e. that g(E(X)) ≤ E(g(X)).

127.

a. $E(X) = 150 + (850 - 150)\dfrac{8}{8+2} = 710$; $V(X) = \dfrac{(850-150)^2(8)(2)}{(8+2)^2(8+2+1)} \rightarrow \sigma_X \approx 84.423$

$P(|X - 710| \leq 84.423) = P(625.577 \leq X \leq 794.423) =$

$\displaystyle\int_{625.577}^{794.423} \dfrac{1}{700}\dfrac{\Gamma(10)}{\Gamma(8)\Gamma(2)}\left(\dfrac{x-150}{700}\right)^7\left(\dfrac{850-x}{700}\right)^1 dx = .684$. The computation of this

integral requires a calculator or computer.

b. $P(X > 750) = \displaystyle\int_{750}^{850} \dfrac{1}{700}\dfrac{\Gamma(10)}{\Gamma(8)\Gamma(2)}\left(\dfrac{x-150}{700}\right)^7\left(\dfrac{850-x}{700}\right)^1 dx = .376$. Again, the

computation of the requested integral requires a calculator or computer.

CHAPTER 5

Section 5.1

1.

    a. $P(X = 1, Y = 1) = p(1,1) = .20$

    b. $P(X \le 1$ and $Y \le 1) = p(0,0) + p(0,1) + p(1,0) + p(1,1) = .42$

    c. At least one hose is in use at both islands. $P(X \ne 0$ and $Y \ne 0) = p(1,1) + p(1,2) + p(2,1) + p(2,2) = .70$

    d. By summing row probabilities, $p_x(x) = .16, .34, .50$ for $x = 0, 1, 2$, and by summing column probabilities, $p_y(y) = .24, .38, .38$ for $y = 0, 1, 2$. $P(X \le 1) = p_x(0) + p_x(1) = .50$

    e. $P(0,0) = .10$, but $p_x(0) \cdot p_y(0) = (.16)(.24) = .0384 \ne .10$, so X and Y are not independent.

3.

    a. $p(1,1) = .15$, the entry in the 1st row and 1st column of the joint probability table.

    b. $P(X_1 = X_2) = p(0,0) + p(1,1) + p(2,2) + p(3,3) = .08 + .15 + .10 + .07 = .40$

    c. $A = \{ (x_1, x_2): x_1 \ge 2 + x_2 \} \cup \{ (x_1, x_2): x_2 \ge 2 + x_1 \}$
       $P(A) = p(2,0) + p(3,0) + p(4,0) + p(3,1) + p(4,1) + p(4,2) + p(0,2) + p(0,3) + p(1,3) = .22$

    d. $P($ exactly $4) = p(1,3) + p(2,2) + p(3,1) + p(4,0) = .17$
       $P($ at least $4) = P($ exactly $4) + p(4,1) + p(4,2) + p(4,3) + p(3,2) + p(3,3) + p(2,3) = .46$

5.

    a. P(X = 3, Y = 3) = P(3 customers, each with 1 package)
                        = P(each has 1 package | 3 customers) · P(3 customers)
                        = $(.6)^3 \cdot (.25) = .054$

    b. P(X = 4, Y = 11) = P(total of 11 packages | 4 customers) · P(4 customers)
        Given that there are 4 customers, there are 4 different ways to have a total of 11
        packages: 3, 3, 3, 2 or 3, 3, 2, 3 or 3, 2, 3 ,3 or 2, 3, 3, 3. Each way has
        probability $(.1)^3(.3)$, so p(4, 11) = $4(.1)^3(.3)(.15)$ = .00018

7.

    a. p(1,1) = .030

    b. P(X ≤ 1 and Y ≤ 1 = p(0,0) + p(0,1) + p(1,0) + p(1,1) = .120

    c. P(X = 1) = p(1,0) + p(1,1) + p(1,2) = .100; P(Y = 1) = p(0,1) + ... + p(5,1) = .300

    d. P(overflow) = P(X + 3Y > 5) = 1 – P(X + 3Y ≤ 5) = 1 – P[(X,Y)=(0,0) or ...or
        (5,0) or (0,1) or (1,1) or (2,1)] = 1 – .620 = .380

    e. The marginal probabilities for X (row sums from the joint probability table) are
        $p_x(0)$ = .05, $p_x(1)$ = .10 , $p_x(2)$ = .25, $p_x(3)$ = .30, $p_x(4)$ = .20, $p_x(5)$ = .10; those
        for Y (column sums) are $p_y(0)$ = .5, $p_y(1)$ = .3, $p_y(2)$ = .2. It is now easily verified
        that for every (x,y), p(x,y) = $p_x(x) \cdot p_y(y)$, so X and Y are independent.

9.

    a. $1 = \int_{-\infty}^{\infty} \int_{-\infty}^{\infty} f(x, y)dxdy = \int_{20}^{30} \int_{20}^{30} K(x^2 + y^2)dxdy$

        $= K \int_{20}^{30} \int_{20}^{30} x^2 \, dydx + K \int_{20}^{30} \int_{20}^{30} y^2 \, dxdy = 10K \int_{20}^{30} x^2 \, dx + 10K \int_{20}^{30} y^2 \, dy$

        $= 20K \cdot \left(\dfrac{19,000}{3}\right) \Rightarrow K = \dfrac{3}{380,000}$

    b. P(X < 26 and Y < 26) = $\int_{20}^{26} \int_{20}^{26} K(x^2 + y^2)dxdy = 12K \int_{20}^{26} x^2 \, dx$

        $= 4Kx^3 \Big|_{20}^{26} = 38,304K = .3024$

c.

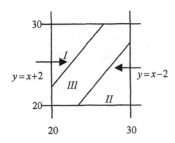

$$P(\,|\,X-Y\,|\le 2\,) = \iint_{\substack{region \\ III}} f(x,y)dxdy = 1 - \iint_{I} f(x,y)dxdy - \iint_{II} f(x,y)dxdy$$

$$= 1 - \int_{20}^{28}\int_{x+2}^{30} f(x,y)dydx - \int_{22}^{30}\int_{20}^{x-2} f(x,y)dydx = .3593 \text{ (after much algebra)}$$

d. $\displaystyle f_x(x) = \int_{-\infty}^{\infty} f(x,y)dy = \int_{20}^{30} K(x^2+y^2)dy = 10Kx^2 + K\frac{y^3}{3}\Big|_{20}^{30}$

$$= 10Kx^2 + .05, \qquad\qquad 20 \le x \le 30$$

e. $f_y(y)$ is obtained by substituting y for x in (d); clearly $f(x,y) \ne f_x(x) \cdot f_y(y)$, so X and Y are not independent.

11.

a. $\displaystyle p(x,y) = \frac{e^{-\lambda}\lambda^x}{x!} \cdot \frac{e^{-\mu}\mu^y}{y!}$ for x = 0, 1, 2, …; y = 0, 1, 2, …

b. $p(0,0) + p(0,1) + p(1,0) = e^{-\lambda-\mu}\left[1 + \lambda + \mu\right]$

c. $\displaystyle P(X+Y=m) = \sum_{k=0}^{m} P(X=k, Y=m-k) = \sum_{k=0}^{m} e^{-\lambda-\mu} \frac{\lambda^k}{k!}\frac{\mu^{m-k}}{(m-k)!}$

$$\frac{e^{-(\lambda+\mu)}}{m!}\sum_{k=0}^{m}\binom{m}{k}\lambda^k \mu^{m-k} = \frac{e^{-(\lambda+\mu)}(\lambda+\mu)^m}{m!}, \text{ so the total \# of errors X+Y also}$$

has a Poisson distribution with parameter $\lambda + \mu$.

13.

a. $f(x,y) = f_x(x) \cdot f_y(y) = \begin{cases} e^{-x-y} & x \geq 0, y \geq 0 \\ 0 & \textit{otherwise} \end{cases}$

b. $P(X \leq 1 \text{ and } Y \leq 1) = P(X \leq 1) \cdot P(Y \leq 1) = (1 - e^{-1})(1 - e^{-1}) = .400$

c. $P(X + Y \leq 2) = \int_0^2 \int_0^{2-x} e^{-x-y} \, dy \, dx = \int_0^2 e^{-x} \left[1 - e^{-(2-x)} \right] dx$

$$= \int_0^2 (e^{-x} - e^{-2}) \, dx = 1 - e^{-2} - 2e^{-2} = .594$$

d. $P(X + Y \leq 1) = \int_0^1 e^{-x} \left[1 - e^{-(1-x)} \right] dx = 1 - 2e^{-1} = .264,$

so $P(1 \leq X + Y \leq 2) = P(X + Y \leq 2) - P(X + Y \leq 1) = .594 - .264 = .330$

15.

a. $F(y) = P(Y \leq y) = P[(X_1 \leq y) \cup ((X_2 \leq y) \cap (X_3 \leq y))]$
$= P(X_1 \leq y) + P[(X_2 \leq y) \cap (X_3 \leq y)] - P[(X_1 \leq y) \cap (X_2 \leq y) \cap (X_3 \leq y)]$
$= (1 - e^{-\lambda y}) + (1 - e^{-\lambda y})^2 - (1 - e^{-\lambda y})^3 \text{ for } y \geq 0$

b. $f(y) = F'(y) = \lambda e^{-\lambda y} + 2(1 - e^{-\lambda y})(\lambda e^{-\lambda y}) - 3(1 - e^{-\lambda y})^2 (\lambda e^{-\lambda y})$
$= 4\lambda e^{-2\lambda y} - 3\lambda e^{-3\lambda y} \text{ for } y \geq 0$

$$E(Y) = \int_0^\infty y \cdot \left(4\lambda e^{-2\lambda y} - 3\lambda e^{-3\lambda y} \right) dy = 2 \left(\frac{1}{2\lambda} \right) - \frac{1}{3\lambda} = \frac{2}{3\lambda}$$

17.

a. $P((X,Y) \text{ within a circle of radius } \frac{R}{2}) = P(A) = \iint_A f(x,y)dxdy$

$$= \frac{1}{\pi R^2} \iint_A dxdy = \frac{area.of.A}{\pi R^2} = \frac{1}{4} = .25$$

b.

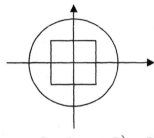

$$P\left(-\frac{R}{2} \le X \le \frac{R}{2}, -\frac{R}{2} \le Y \le \frac{R}{2}\right) = \frac{R^2}{\pi R^2} = \frac{1}{\pi}$$

c.

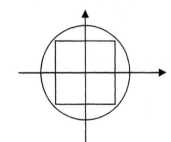

$$P\left(-\frac{R}{\sqrt{2}} \le X \le \frac{R}{\sqrt{2}}, -\frac{R}{\sqrt{2}} \le Y \le \frac{R}{\sqrt{2}}\right) = \frac{2R^2}{\pi R^2} = \frac{2}{\pi}$$

d. $f_x(x) = \int_{-\infty}^{\infty} f(x,y)dy = \int_{-\sqrt{R^2-x^2}}^{\sqrt{R^2-x^2}} \frac{1}{\pi R^2} dy = \frac{2\sqrt{R^2-x^2}}{\pi R^2}$ for $-R \le x \le R$ and

similarly for $f_Y(y)$. X and Y are <u>not</u> independent since $f_X(x)f_Y(y) \ne f(x,y)$.

19.

   a. $f_{Y|X}(y \mid x) = \dfrac{f(x,y)}{f_X(x)} = \dfrac{k(x^2 + y^2)}{10kx^2 + .05}$ for $20 \le y \le 30$

      $f_{X|Y}(x \mid y) = \dfrac{k(x^2 + y^2)}{10ky^2 + .05}$ for $20 \le x \le 30$ $\left(k = \dfrac{3}{380,000} \right)$

   b. $P(Y \ge 25 \mid X = 22) = \displaystyle\int_{25}^{30} f_{Y|X}(y \mid 22)\, dy$

                 $= \displaystyle\int_{25}^{30} \dfrac{k((22)^2 + y^2)}{10k(22)^2 + .05}\, dy = .783$

      $P(Y \ge 25) = \displaystyle\int_{25}^{30} f_Y(y)\, dy = \int_{25}^{30}(10ky^2 + .05)\, dy = .75$

   c. $E(Y \mid X = 22) = \displaystyle\int_{-\infty}^{\infty} y \cdot f_{Y|X}(y \mid 22)\, dy = \int_{20}^{30} y \cdot \dfrac{k((22)^2 + y^2)}{10k(22)^2 + .05}\, dy = 25.372912$

      $E(Y^2 \mid X = 22) = \displaystyle\int_{20}^{30} y^2 \cdot \dfrac{k((22)^2 + y^2)}{10k(22)^2 + .05}\, dy = 652.028640$

      $V(Y \mid X = 22) = E(Y^2 \mid X = 22) - [E(Y \mid X = 22)]^2 = 8.243976$

21.

   a. $f_{x_3 | x_1, x_2}(x_3 \mid x_1, x_2) = \dfrac{f(x_1, x_2, x_3)}{f_{x_1, x_2}(x_1, x_2)}$, where $f_{x_1, x_2}(x_1, x_2) = $ the marginal joint

      pdf of $(X_1, X_2) = \displaystyle\int_{-\infty}^{\infty} f(x_1, x_2, x_3)\, dx_3$

   b. $f_{x_2, x_3 | x_1}(x_2, x_3 \mid x_1) = \dfrac{f(x_1, x_2, x_3)}{f_{x_1}(x_1)}$, where

      $f_{x_1}(x_1) = \displaystyle\int_{-\infty}^{\infty} \int_{-\infty}^{\infty} f(x_1, x_2, x_3)\, dx_2\, dx_3$

Section 5.2

23. $E(X_1 - X_2) = \displaystyle\sum_{x_1=0}^{4} \sum_{x_2=0}^{3}(x_1 - x_2)\cdot p(x_1, x_2) = (0-0)(.08) + (0-1)(.07) + \ldots +$

$(4-3)(.06) = .15$; this also equals $E(X_1) - E(X_2) = 1.70 - 1.55$

25. $E(XY) = E(X) \cdot E(Y) = L \cdot L = L^2$

27. $\displaystyle\int_0^1 \int_0^1 |x - y| \cdot f(x,y)\,dxdy = \int_0^1 \int_0^1 |x - y| \cdot 6x^2 y\,dxdy$

$= \displaystyle\int_0^1 \int_0^x (x - y) \cdot 6x^2 y\,dydx + \int_0^1 \int_x^1 (x - y) \cdot 6x^2 y\,dydx = \frac{1}{6} + \frac{1}{12} = \frac{1}{4}$

29. $\text{Cov(X,Y)} = -\dfrac{2}{75}$ and $\mu_x = \mu_y = \dfrac{2}{5}$. $E(X^2) = \displaystyle\int_0^1 x^2 \cdot f_x(x)\,dx$

$= 12\displaystyle\int_0^1 x^3(1 - x^2\,dx) = \dfrac{12}{60} = \dfrac{1}{5}$, so Var (X) $= \dfrac{1}{5} - \dfrac{4}{25} = \dfrac{1}{25}$

Similarly, Var(Y) $= \dfrac{1}{25}$, so $\rho_{X,Y} = \dfrac{-2/75}{\sqrt{\frac{1}{25}} \cdot \sqrt{\frac{1}{25}}} = -\dfrac{50}{75} = -.667$

31.

a. $E(X) = \displaystyle\int_{20}^{30} xf_x(x)\,dx = \int_{20}^{30} x\left[10Kx^2 + .05\right]dx = 25.329 = E(Y)$

$E(XY) = \displaystyle\int_{20}^{30} \int_{20}^{30} xy \cdot K(x^2 + y^2)\,dxdy = 641.447$

$\Rightarrow Cov(X,Y) = 641.447 - (25.329)^2 = -.111$

b. $E(X^2) = \displaystyle\int_{20}^{30} x^2\left[10Kx^2 + .05\right]dx = 649.8246 = E(Y^2)$,

so Var (X) = Var(Y) $= 649.8246 - (25.329)^2 = 8.2664$

$\Rightarrow \rho = \dfrac{-.111}{\sqrt{(8.2664)(8.2664)}} = -.0134$

33. Since $E(XY) = E(X) \cdot E(Y)$, $Cov(X,Y) = E(XY) - E(X) \cdot E(Y) = E(X) \cdot E(Y) - E(X) \cdot$ $E(Y) = 0$, and since $Corr(X,Y) = \dfrac{Cov(X,Y)}{\sigma_x \sigma_y}$, then $Corr(X,Y) = 0$

35.

a. $Cov(aX + b, cY + d) = E[(aX + b)(cY + d)] - E(aX + b) \cdot E(cY + d)$
$= E[acXY + adX + bcY + bd] - (aE(X) + b)(cE(Y) + d)$
$= acE(XY) - acE(X)E(Y) = acCov(X,Y)$

b. $Corr(aX + b, cY + d) =$
$$\frac{Cov(aX + b, cY + d)}{\sqrt{Var(aX + b)}\sqrt{Var(cY + d)}} = \frac{acCov(X,Y)}{|a| \cdot |c| \sqrt{Var(X) \cdot Var(Y)}} = Corr(X,Y) \text{ when a}$$
and c have the same signs.

When a and c differ in sign, $Corr(aX + b, cY + d) = -Corr(X,Y)$.

Section 5.3

37.

$P(x_2)$	$P(x_1)$ $x_2 \mid x_1$	.20 25	.50 40	.30 65
.20	25	.04	.10	.06
.50	40	.10	.25	.15
.30	65	.06	.15	.09

a.

$\bar{x}$	25	32.5	40	45	52.5	65
$p(\bar{x})$	.04	.20	.25	.12	.30	.09

$E(\bar{x}) = (25)(.04) + 32.5(.20) + \ldots + 65(.09) = 44.5 = \mu$

b.

s^2	0	112.5	312.5	800
$p(s^2)$	.38	.20	.30	.12

$E(s^2) = 212.25 = \sigma^2$

39.

x	0	1	2	3	4	5	6	7	8	9	10
x/n	0	.1	.2	.3	.4	.5	.6	.7	.8	.9	1.0
p(x/n)	.000	.000	.000	.001	.005	.027	.088	.201	.302	.269	.107

X is a binomial random variable with p = .8.

41.

Outcome	1,1	1,2	1,3	1,4	2,1	2,2	2,3	2,4
Probability	.16	.12	.08	.04	.12	.09	.06	.03
$\overline{x}$	1	1.5	2	2.5	1.5	2	2.5	3
r	0	1	2	3	1	0	1	2

Outcome	3,1	3,2	3,3	3,4	4,1	4,2	4,3	4,4
Probability	.08	.06	.04	.02	.04	.03	.02	.01
$\overline{x}$	2	2.5	3	3.5	2.5	3	3.5	4
r	2	1	0	1	3	2	1	2

a.

$\overline{x}$	1	1.5	2	2.5	3	3.5	4
$p(\overline{x})$	.16	.24	.25	.20	.10	.04	.01

b. $P(\overline{x} \le 2.5) = .8$

c.

r	0	1	2	3
p(r)	.30	.40	.22	.08

d. $P(\overline{X} \le 1.5) = P(1,1,1,1) + P(2,1,1,1) + \ldots + P(1,1,1,2) + P(1,1,2,2) + \ldots + P(2,2,1,1) + P(3,1,1,1) + \ldots + P(1,1,1,3) = (.4)^4 + 4(.4)^3(.3) + 6(.4)^2(.3)^2 + 4(.4)^2(.2)^2 = .2400$

Chapter 5: Joint Probability Distributions and Random Samples

43. The statistic of interest is the fourth spread, or the difference between the medians of the upper and lower halves of the data. The population distribution is uniform with A = 8 and B = 10. Use a computer to generate samples of sizes n = 5, 10, 20, and 30 from a uniform distribution with A = 8 and B = 10. Keep the number of replications the same (say 500, for example). For each sample, compute the upper and lower fourth, then compute the difference. Plot the sampling distributions on separate histograms for n = 5, 10, 20, and 30.

45. Using Minitab to generate the necessary sampling distribution, we can see that as n increases, the distribution slowly moves toward normality. However, even the sampling distribution for n = 50 is not yet approximately normal.

n = 10

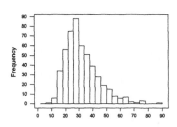

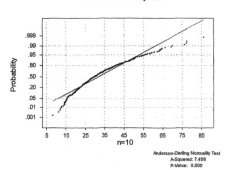

n = 50

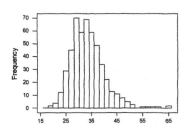

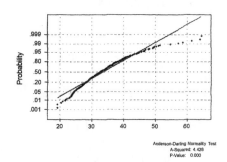

Section 5.4

47. $\mu = 12$ cm, $\sigma = .04$ cm

   a. $n = 16$: $P(11.99 \leq \overline{X} \leq 12.01) = P\left(\dfrac{11.99-12}{.01} \leq Z \leq \dfrac{12.01-12}{.01}\right)$

   $= P(-1 \leq Z \leq 1) = \Phi(1) - \Phi(-1) = .8413 - .1587 = .6826$

   b. $n = 25$: $P(\overline{X} > 12.01) = P\left(Z > \dfrac{12.01-12}{.04/5}\right) = P(Z > 1.25)$

   $= 1 - \Phi(1.25) = 1 - .8944 = .1056$

49.

   a. 11 P.M. – 6:50 P.M. = 250 minutes. With $T_0 = X_1 + \ldots + X_{40}$ = total grading time, $\mu_{T_0} = n\mu = (40)(6) = 240$ and $\sigma_{T_0} = \sigma\sqrt{n} = 37.95$, so $P(T_0 \leq 250) \approx$

   $P\left(Z \leq \dfrac{250-240}{37.95}\right) = P(Z \leq .26) = .6026$

   b. $P(T_0 > 260) = P\left(Z > \dfrac{260-240}{37.95}\right) = P(Z > .53) = .2981$

51. $X \sim N(10,4)$.

   For day 1, $n = 5$: $P(\overline{X} \leq 11) = P\left(Z \leq \dfrac{11-10}{2/\sqrt{5}}\right) = P(Z \leq 1.12) = .8686$

   For day 2, $n = 6$: $P(\overline{X} \leq 11) = P\left(Z \leq \dfrac{11-10}{2/\sqrt{6}}\right) = P(Z \leq 1.22) = .8888$

   For both days, $(.8686)(.8888) = .7720$

53. $\mu = 50$, $\sigma = 1.2$

   a.   n = 9

$$P(\overline{X} \geq 51) = P\left(Z \geq \frac{51-50}{1.2/\sqrt{9}} \right) = P(Z \geq 2.5) = 1 - .9938 = .0062$$

   b.   n = 40

$$P(\overline{X} \geq 51) = P\left(Z \geq \frac{51-50}{1.2/\sqrt{40}} \right) = P(Z \geq 5.27) \approx 0$$

55.

   a.   With Y = # of tickets, Y has approximately a normal distribution with
$\mu = \lambda = 50$, $\sigma = \sqrt{\lambda} = 7.071$, so $P(35 \leq Y \leq 70) \approx$

$$P\left(\frac{34.5-50}{7.071} \leq Z \leq \frac{70.5-50}{7.071} \right) = P(-2.19 \leq Z \leq 2.90) = .9838$$

   b.   Here $\mu = 250$, $\sigma^2 = 250, \sigma = 15.811$, so $P(225 \leq Y \leq 275) \approx$

$$P\left(\frac{224.5-250}{15.811} \leq Z \leq \frac{275.5-250}{15.811} \right) = P(-1.61 \leq Z \leq 1.61) = .8926$$

57.   E(X) = 100, Var(X) = 200, $\sigma_x = 14.14$, so $P(X \leq 125) \approx P\left(Z \leq \frac{125-100}{14.14} \right)$

$$= P(Z \leq 1.77) = .9616$$

Section 5.5

59.

a. $E(X_1 + X_2 + X_3) = 180$, $V(X_1 + X_2 + X_3) = 45$, $\sigma_{x_1+x_2+x_3} = 6.708$

$$P(X_1 + X_2 + X_3 \leq 200) = P\left(Z \leq \frac{200-180}{6.708}\right) = P(Z \leq 2.98) = .9986$$

$$P(150 \leq X_1 + X_2 + X_3 \leq 200) = P(-4.47 \leq Z \leq 2.98) \approx .9986$$

b. $\mu_{\overline{X}} = \mu = 60$, $\sigma_{\overline{x}} = \dfrac{\sigma_x}{\sqrt{n}} = \dfrac{\sqrt{15}}{\sqrt{3}} = 2.236$

$$P(\overline{X} \geq 55) = P\left(Z \geq \frac{55-60}{2.236}\right) = P(Z \geq -2.236) = .9875$$

$$P(58 \leq \overline{X} \leq 62) = P(-.89 \leq Z \leq .89) = .6266$$

c. $E(X_1 - .5X_2 - .5X_3) = 0$;

$V(X_1 - .5X_2 - .5X_3) = \sigma_1^2 + .25\sigma_2^2 + .25\sigma_3^2 = 22.5 \rightarrow$ sd $= 4.7434$

$$P(-10 \leq X_1 - .5X_2 - .5X_3 \leq 5) = P\left(\frac{-10-0}{4.7434} \leq Z \leq \frac{5-0}{4.7434}\right)$$

$$= P(-2.11 \leq Z \leq 1.05) = .8531 - .0174 = .8357$$

d. $E(X_1 + X_2 + X_3) = 150$, $V(X_1 + X_2 + X_3) = 36$, $\sigma_{x_1+x_2+x_3} = 6$

$$P(X_1 + X_2 + X_3 \leq 200) = P\left(Z \leq \frac{160-150}{6}\right) = P(Z \leq 1.67) = .9525$$

We want $P(X_1 + X_2 \geq 2X_3)$, or written another way, $P(X_1 + X_2 - 2X_3 \geq 0)$
$E(X_1 + X_2 - 2X_3) = 40 + 50 - 2(60) = -30$,
$V(X_1 + X_2 - 2X_3) = \sigma_1^2 + \sigma_2^2 + 4\sigma_3^2 = 78 \rightarrow$ sd $= 8.832$, so

$$P(X_1 + X_2 - 2X_3 \geq 0) = P\left(Z \geq \frac{0-(-30)}{8.832}\right) = P(Z \geq 3.40) = .0003$$

61.

a. The marginal pmf's of X and Y are given in the solution to Exercise 7, from which $E(X) = 2.8$, $E(Y) = .7$, $V(X) = 1.66$, $V(Y) = .61$. Thus $E(X+Y) = E(X) + E(Y) = 3.5$, $V(X+Y) = V(X) + V(Y) = 2.27$, and the standard deviation of $X + Y$ is 1.51

b. $E(3X+10Y) = 3E(X) + 10E(Y) = 15.4$, $V(3X+10Y) = 9V(X) + 100V(Y) = 75.94$, and the standard deviation of revenue is 8.71

Chapter 5: Joint Probability Distributions and Random Samples

63.

    a. $E(X_1) = 1.70$, $E(X_2) = 1.55$, $E(X_1X_2) = \sum_{x_1} \sum_{x_2} x_1 x_2 \, p(x_1, x_2) = 3.33$, so

        $Cov(X_1,X_2) = E(X_1X_2) - E(X_1) \, E(X_2) = 3.33 - 2.635 = .695$

    b. $V(X_1 + X_2) = V(X_1) + V(X_2) + 2 \, Cov(X_1,X_2) = 1.59 + 1.0875 + 2(.695) = 4.0675$.
        This is much larger than $V(X_1) + V(X_2)$, since the two variables are positively
        correlated.

65. $\mu = 5.00$, $\sigma = .2$

    a. $E(\overline{X} - \overline{Y}) = 0$; $V(\overline{X} - \overline{Y}) = \dfrac{\sigma^2}{25} + \dfrac{\sigma^2}{25} = .0032$, $\sigma_{\overline{X}-\overline{Y}} = .0566$

        $\Rightarrow P\!\left(-.1 \le \overline{X} - \overline{Y} \le .1\right) = P\!\left(-1.77 \le Z \le 1.77\right) = .9232$

    b. $V(\overline{X} - \overline{Y}) = \dfrac{\sigma^2}{36} + \dfrac{\sigma^2}{36} = .0022222$, $\sigma_{\overline{X}-\overline{Y}} = .0471$
        $\Rightarrow P\!\left(-.1 \le \overline{X} - \overline{Y} \le .1\right) \approx P\!\left(-2.12 \le Z \le 2.12\right) = .9660$ (by the CLT)

67.   Letting X_1, X_2, and X_3 denote the lengths of the three pieces, the total length is
    $X_1 + X_2 - X_3$. This has a normal distribution with mean value $20 + 15 - 1 = 34$,
    variance $.25+.16+.01 = .42$, and standard deviation $.6481$. Standardizing gives
    $P(34.5 \le X_1 + X_2 - X_3 \le 35) = P(.77 \le Z \le 1.54) = .1588$

69.

    a. $E(X_1 + X_2 + X_3) = 800 + 1000 + 600 = 2400$.

    b. Assuming independence of X_1, X_2, X_3, $Var(X_1 + X_2 + X_3)$
        $= (16)^2 + (25)^2 + (18)^2 = 1205$

    c. $E(X_1 + X_2 + X_3) = 2400$ as before, but now $Var(X_1 + X_2 + X_3)$
        $= Var(X_1) + Var(X_2) + Var(X_3) + 2Cov(X_1,X_2) + 2Cov(X_1, X_3) + 2Cov(X_2, X_3) =$
        1745, with sd $= 41.77$

71.

a. $M = a_1 X_1 + a_2 X_2 + W \int_0^{12} x\,dx = a_1 X_1 + a_2 X_2 + 72W$, so

$E(M) = (5)(2) + (10)(4) + (72)(1.5) = 158m$

$\sigma_M^2 = (5)^2(.5)^2 + (10)^2(1)^2 + (72)^2(.25)^2 = 430.25$, $\sigma_M = 20.74$

b. $P(M \le 200) = P\left(Z \le \dfrac{200-158}{20.74}\right) = P(Z \le 2.03) = .9788$

73.

a. Both approximately normal by the C.L.T.

b. The difference of two r.v.'s is just a special linear combination, and a linear combination of normal r.v's has a normal distribution, so $\overline{X} - \overline{Y}$ has approximately a normal distribution with $\mu_{\overline{X}-\overline{Y}} = 5$ and

$\sigma_{\overline{X}-\overline{Y}}^2 = \dfrac{8^2}{40} + \dfrac{6^2}{35} = 2.629, \sigma_{\overline{X}-\overline{Y}} = 1.621$

c. $P\left(-1 \le \overline{X} - \overline{Y} \le 1\right) \approx P\left(\dfrac{-1-5}{1.6213} \le Z \le \dfrac{1-5}{1.6213}\right) = P(-3.70 \le Z \le -2.47) \approx .0068$

d. $P\left(\overline{X} - \overline{Y} \ge 10\right) \approx P\left(Z \ge \dfrac{10-5}{1.6213}\right) = P(Z \ge 3.08) = .0010$. This probability is quite

small, so such an occurrence is unlikely if $\mu_1 - \mu_2 = 5$, and we would thus doubt this claim.

Supplementary Exercises

75.

   a. $p_X(x)$ is obtained by adding joint probabilities across the row labeled x, resulting in $p_X(x) = .2, .5, .3$ for $x = 12, 15, 20$ respectively. Similarly, from column sums $p_Y(y) = .1, .35, .55$ for $y = 12, 15, 20$ respectively.

   b. $P(X \le 15 \text{ and } Y \le 15) = p(12,12) + p(12,15) + p(15,12) + p(15,15) = .25$

   c. $p_x(12) \cdot p_y(12) = (.2)(.1) \ne .05 = p(12,12)$, so X and Y are not independent. (Almost any other (x,y) pair yields the same conclusion).

   d. $E(X + Y) = \sum\sum (x + y)p(x, y) = 33.35$ (or $= E(X) + E(Y) = 33.35$)

   e. $E(|X - Y|) = \sum\sum |x - y| p(x, y) = 3.85$

77.

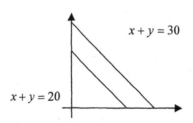

$x + y = 30$

$x + y = 20$

   a. $1 = \int_{-\infty}^{\infty}\int_{-\infty}^{\infty} f(x, y)\,dx\,dy = \int_0^{20}\int_{20-x}^{30-x} kxy\,dy\,dx + \int_{20}^{30}\int_0^{30-x} kxy\,dy\,dx$

$= \dfrac{81,250}{3} \cdot k \Rightarrow k = \dfrac{3}{81,250}$

   b. $f_X(x) = \begin{cases} \int_{20-x}^{30-x} kxy\,dy = k(250x - 10x^2) & 0 \le x \le 20 \\ \int_0^{30-x} kxy\,dy = k(450x - 30x^2 + \tfrac{1}{2}x^3) & 20 \le x \le 30 \end{cases}$

By symmetry, $f_Y(y)$ is obtained by substituting y for x in $f_X(x)$. Since $f_X(25) > 0$, and $f_Y(25) > 0$, but $f(25, 25) = 0$, $f_X(x) \cdot f_Y(y) \ne f(x,y)$ for all x,y. X and Y are not independent.

c. $P(X+Y \leq 25) = \int_0^{20} \int_{20-x}^{25-x} kxy \, dy \, dx + \int_{20}^{25} \int_0^{25-x} kxy \, dy \, dx = \dfrac{3}{81,250} \cdot \dfrac{230,625}{24} = .355$

d. $E(X+Y) = E(X) + E(Y) = 2\left\{ \int_0^{20} x \cdot k\left(250x - 10x^2\right) dx \right.$

$\left. + \int_{20}^{30} x \cdot k\left(450x - 30x^2 + \tfrac{1}{2}x^3\right) dx \right\} = 2k(351,666.67) = 25.969$

e. $E(XY) = \int_{-\infty}^{\infty} \int_{-\infty}^{\infty} xy \cdot f(x,y) \, dx \, dy = \int_0^{20} \int_{20-x}^{30-x} kx^2 y^2 \, dy \, dx$

$+ \int_{20}^{30} \int_0^{30-x} kx^2 y^2 \, dy \, dx = \dfrac{k}{3} \cdot \dfrac{33,250,000}{3} = 136.4103$, so

$\mathrm{Cov}(X,Y) = 136.4103 - (12.9845)^2 = -32.19$, and $E(X^2) = E(Y^2) = 204.6154$, so

$\sigma_x^2 = \sigma_y^2 = 204.6154 - (12.9845)^2 = 36.0182$ and $\rho = \dfrac{-32.19}{36.0182} = -.894$

f. $\mathrm{Var}(X+Y) = \mathrm{Var}(X) + \mathrm{Var}(Y) + 2\mathrm{Cov}(X,Y) = 7.66$

79. $E(\overline{X} + \overline{Y} + \overline{Z}) = 500 + 900 + 2000 = 3400$

$V(\overline{X} + \overline{Y} + \overline{Z}) = \dfrac{50^2}{365} + \dfrac{100^2}{365} + \dfrac{180^2}{365} = 123.014 \rightarrow sd = 11.09.$

$P(\overline{X} + \overline{Y} + \overline{Z} \leq 3500) = P(Z \leq 9.0) \approx 1$

81.

a. $E(N) \cdot \mu = (10)(40) = 400$ minutes

b. We expect 20 components to come in for repair during a 4 hour period,
so $E(N) \cdot \mu = (20)(3.5) = 70$

83. $0.95 = P(\mu - .02 \leq \overline{X} \leq \mu + .02) \approx P\left(\dfrac{-.02}{.1/\sqrt{n}} \leq Z \leq \dfrac{.02}{.1/\sqrt{n}} \right) = P\left(-.2\sqrt{n} \leq Z \leq .2\sqrt{n} \right)$

but $P(-1.96 \leq Z \leq 1.96) = .95$ so $.2\sqrt{n} = 1.96 \Rightarrow n = 97$. The C.L.T.

85.   The expected value and standard deviation of volume are 87,850 and 4370.37, respectively, so

$$P(volume \le 100,000) = P\left(Z \le \frac{100,000 - 87,850}{4370.37}\right) = P(Z \le 2.78) = .9973$$

87.

a. $Var(aX + Y) = a^2\sigma_x^2 + 2aCov(X,Y) + \sigma_y^2 = a^2\sigma_x^2 + 2a\sigma_X\sigma_Y\rho + \sigma_y^2$.

Substituting $a = \dfrac{\sigma_Y}{\sigma_X}$ yields $\sigma_Y^2 + 2\sigma_Y^2\rho + \sigma_Y^2 = 2\sigma_Y^2(1-\rho) \ge 0$, so $\rho \ge -1$

b.   Same argument as in **a**

c.   Suppose $\rho = 1$. Then $Var(aX - Y) = 2\sigma_Y^2(1-\rho) = 0$, which implies that $aX - Y = k$ (a constant), so $Y = aX - k$, which is of the form $aX + b$.

89.

a.   With Y = X$_1$ + X$_2$,

$$F_Y(y) = \int_0^y \left\{ \int_0^{y-x_1} \frac{1}{2^{\nu_1/2}\Gamma(\nu_1/2)} \cdot \frac{1}{2^{\nu_{21}/2}\Gamma(\nu_2/2)} \cdot x_1^{\frac{\nu_1}{2}-1} x_2^{\frac{\nu_2}{2}-1} e^{-\frac{x_1+x_2}{2}} dx_2 \right\} dx_1.$$

But the inner integral can be shown to be equal to

$$\frac{1}{2^{(\nu_1+\nu_2)/2}\Gamma((\nu_1+\nu_2)/2)} y^{[(\nu_1+\nu_2)/2]-1} e^{-y/2},$$ from which the result follows.

b.   By **a**, $Z_1^2 + Z_2^2$ is chi-squared with $v = 2$, so $\left(Z_1^2 + Z_2^2\right) + Z_3^2$ is chi-squared with $v = 3$, etc., until $Z_1^2 + ... + Z_n^2$ is chi-squared with $v = n$

c. $\dfrac{X_i - \mu}{\sigma}$ is standard normal, so $\left[\dfrac{X_i - \mu}{\sigma}\right]^2$ is chi-squared with $v = 1$, so the sum is chi-squared with $v = n$

Chapter 5: Joint Probability Distributions and Random Samples

91.

a. $V(X_1) = V(W + E_1) = \sigma_W^2 + \sigma_E^2 = V(W + E_2) = V(X_2)$ and
$Cov(X_1, X_2) = Cov(W + E_1, W + E_2) = Cov(W, W) + Cov(W, E_2) +$
$Cov(E_1, W) + Cov(E_1, E_2) = Cov(W, W) = V(W) = \sigma_w^2$.

Thus, $\rho = \dfrac{\sigma_W^2}{\sqrt{\sigma_W^2 + \sigma_E^2} \cdot \sqrt{\sigma_W^2 + \sigma_E^2}} = \dfrac{\sigma_W^2}{\sigma_W^2 + \sigma_E^2}$

b. $\rho = \dfrac{1}{1 + .0001} = .9999$

93. $E(Y) \approx h(\mu_1, \mu_2, \mu_3, \mu_4) = 120\left[\frac{1}{10} + \frac{1}{15} + \frac{1}{20}\right] = 26$

The partial derivatives of $h(\mu_1, \mu_2, \mu_3, \mu_4)$ with respect to x_1, x_2, x_3, and x_4 are

$-\dfrac{x_4}{x_1^2}$, $-\dfrac{x_4}{x_2^2}$, $-\dfrac{x_4}{x_3^2}$, and $\dfrac{1}{x_1} + \dfrac{1}{x_2} + \dfrac{1}{x_3}$, respectively.

Substituting $x_1 = 10$, $x_2 = 15$, $x_3 = 20$, and $x_4 = 120$ gives -1.2, $-.5333$, $-.3000$, and .2167, respectively.
$V(Y) = (1)(-1.2)^2 + (1)(-.5333)^2 + (1.5)(-.3000)^2 + (4.0)(.2167)^2 = 2.6783$, and the approximate sd of Y is 1.64.

95. Since X and Y are standard normal, each has mean 0 and variance 1.

a. $Cov(X, U) = Cov(X, .6X + .8Y) = .6Cov(X, X) + .8Cov(X, Y) = .6V(X) + 0 = .6(1)$.
The covariance of X and Y is zero because X and Y are independent.
Also, $Var(U) = Var(.6X + .8Y) = (.6)^2 V(X) + (.8)^2 V(Y) = (.36)(1) + (.64)(1) = 1$.
Therefore,

$$Corr(X, U) = \frac{Cov(X, U)}{\sigma_X \sigma_U} = \frac{.6}{\sqrt{1}\sqrt{1}} = .6, \text{ the coefficient on } X$$

b. Based on part **a**, for any specified ρ we want $U = \rho X + bY$, where the coefficient b on Y has the feature that $\rho^2 + b^2 = 1$ (so that the variance of U equals 1). One possible option for b is $b = \sqrt{1 - \rho^2}$, from which $U = \rho X + \sqrt{1 - \rho^2}\, Y$.

CHAPTER 6

Section 6.1

1.

    a. We use the sample mean, $\bar{x}$ to estimate the population mean, μ:

$$\hat{\mu} = \bar{x} = \frac{\Sigma x_i}{n} = \frac{219.80}{27} = 8.1407$$

    b. We use the sample median, $\tilde{x} = 7.7$ (the middle observation when arranged in ascending order)

    c. We use the sample standard deviation, $s = \sqrt{s^2} = \sqrt{\dfrac{1860.94 - \frac{(219.8)^2}{27}}{26}} = 1.660$

    d. With "success" = observation greater than 10, x = # of successes = 4, and

$$\hat{p} = \frac{x}{n} = \frac{4}{27} = .1481$$

    e. We use $\dfrac{s}{\bar{x}} = \dfrac{1.660}{8.1407} = .2039$

3.

    a. We use the sample mean, $\bar{x} = 1.3481$

    b. Because we assume normality, the mean = median, so we also use the sample mean $\bar{x} = 1.3481$. We could also easily use the sample median.

    c. We use the 90th percentile of the sample:
$\hat{\mu} + (1.28)\hat{\sigma} = \bar{x} + 1.28s = 1.3481 + (1.28)(.3385) = 1.7814$.

    d. Since we can assume normality,
$$P(X < 1.5) \approx P\left(Z < \frac{1.5 - \bar{x}}{s}\right) = P\left(Z < \frac{1.5 - 1.3481}{.3385}\right) = P(Z < .45) = .6736$$

    e. The estimated standard error of $\bar{x} = \dfrac{\hat{\sigma}}{\sqrt{n}} = \dfrac{s}{\sqrt{n}} = \dfrac{.3385}{\sqrt{16}} = .0846$

5.   Let θ = the total audited value. Three potential estimators of θ are $\hat{\theta}_1 = N\overline{X}$,

$\hat{\theta}_2 = T - N\overline{D}$, and $\hat{\theta}_3 = T \cdot \dfrac{\overline{X}}{\overline{Y}}$. From the data, $\overline{y} = 374.6$, $\overline{x} = 340.6$, and $\overline{d} = 34.0$.

Knowing $N = 5{,}000$ and $T = 1{,}761{,}300$, the three corresponding estimates

are $\hat{\theta}_1 = (5{,}000)(340.6) = 1{,}703{,}000$, $\hat{\theta}_2 = 1{,}761{,}300 - (5{,}000)(34.0) = 1{,}591{,}300$, and

$\hat{\theta}_3 = 1{,}761{,}300 \left(\dfrac{340.6}{374.6} \right) = 1{,}601{,}438.281$.

7.

a. $\hat{\mu} = \overline{x} = \dfrac{\sum x_i}{n} = \dfrac{1206}{10} = 120.6$

b. $\hat{\tau} = 10{,}000$, $\hat{\mu} = 1{,}206{,}000$

c.   8 of 10 houses in the sample used at least 100 therms (the "successes"), so
      $\hat{p} = \frac{8}{10} = .80$.

d.   The ordered sample values are 89, 99, 103, 109, 118, 122, 125, 138, 147, 156,
      from which the two middle values are 118 and 122, so $\hat{\tilde{\mu}} = \tilde{x} = \dfrac{118 + 122}{2} = 120.0$

9.

a. $E(\overline{X}) = \mu = E(X) = \lambda$, so $\overline{X}$ is an unbiased estimator for the Poisson parameter
      λ; $\sum x_i = (0)(18) + (1)(37) + \ldots + (7)(1) = 317$, since n $= 150$, $\hat{\lambda} = \overline{x} = \dfrac{317}{150} = 2.11$

b. $\sigma_{\overline{x}} = \dfrac{\sigma}{\sqrt{n}} = \dfrac{\sqrt{\lambda}}{\sqrt{n}}$, so the estimated standard error is $\sqrt{\dfrac{\hat{\lambda}}{n}} = \dfrac{\sqrt{2.11}}{\sqrt{150}} = .119$

11.

a. $E\left(\dfrac{X_1}{n_1}-\dfrac{X_2}{n_2}\right)=\dfrac{1}{n_1}E(X_1)-\dfrac{1}{n_2}E(X_2)=\dfrac{1}{n_1}(n_1 p_1)-\dfrac{1}{n_2}(n_2 p_2)=p_1-p_2$.

b. $Var\left(\dfrac{X_1}{n_1}-\dfrac{X_2}{n_2}\right)=Var\left(\dfrac{X_1}{n_1}\right)+Var\left(\dfrac{X_2}{n_2}\right)=\left(\dfrac{1}{n_1}\right)^2 Var(X_1)+\left(\dfrac{1}{n_2}\right)^2 Var(X_2)$

$\dfrac{1}{n_1^2}(n_1 p_1 q_1)+\dfrac{1}{n_2^2}(n_2 p_2 q_2)=\dfrac{p_1 q_1}{n_1}+\dfrac{p_2 q_2}{n_2}$, and the standard error is the square

root of this quantity.

c. With $\hat{p}_1=\dfrac{x_1}{n_1}$, $\hat{q}_1=1-\hat{p}_1$, $\hat{p}_2=\dfrac{x_2}{n_2}$, $\hat{q}_2=1-\hat{p}_2$, the estimated standard error

is $\sqrt{\dfrac{\hat{p}_1 \hat{q}_1}{n_1}+\dfrac{\hat{p}_2 \hat{q}_2}{n_2}}$.

d. $(\hat{p}_1-\hat{p}_2)=\dfrac{127}{200}-\dfrac{176}{200}=.635-.880=-.245$

e. $\sqrt{\dfrac{(.635)(.365)}{200}+\dfrac{(.880)(.120)}{200}}=.041$

13. $E(X)=\displaystyle\int_{-1}^{1}x\cdot\tfrac{1}{2}(1+\theta x)dx=\dfrac{x^2}{4}+\dfrac{\theta x^3}{6}\Big|_{-1}^{1}=\dfrac{1}{3}\theta$, so $E(\overline{X})=\dfrac{1}{3}\theta$. Hence,

$\hat{\theta}=3\overline{X}\Rightarrow E(\hat{\theta})=E(3\overline{X})=3E(\overline{X})=3\left(\dfrac{1}{3}\right)\theta=\theta$

Chapter 6: Point Estimation

15.

a. $E(X^2) = 2\theta$ implies that $E\left(\dfrac{X^2}{2}\right) = \theta$. Consider $\hat{\theta} = \dfrac{\sum X_i^2}{2n}$. Then

$$E(\hat{\theta}) = E\left(\frac{\sum X_i^2}{2n}\right) = \frac{\sum E(X_i^2)}{2n} = \frac{\sum 2\theta}{2n} = \frac{2n\theta}{2n} = \theta \text{, implying that } \hat{\theta} \text{ is an}$$

unbiased estimator for θ.

b. $\sum x_i^2 = 1490.1058$, so $\hat{\theta} = \dfrac{1490.1058}{20} = 74.505$

17.

a. $E(\hat{p}) = \displaystyle\sum_{x=0}^{\infty} \frac{r-1}{x+r-1} \cdot \binom{x+r-1}{x} \cdot p^r \cdot (1-p)^x$

$= p \displaystyle\sum_{x=0}^{\infty} \frac{(x+r-2)!}{x!(r-2)!} \cdot p^{r-1} \cdot (1-p)^x = p \displaystyle\sum_{x=0}^{\infty} \binom{x+r-2}{x} p^{r-1}(1-p)^x$

$= p \displaystyle\sum_{x=0}^{\infty} nb(x; r-1, p) = p$.

b. For the given sequence, $x = 5$, so $\hat{p} = \dfrac{5-1}{5+5-1} = \dfrac{4}{9} = .444$

19.

a. $\lambda = .5p + .15 \Rightarrow 2\lambda = p + .3$, so $p = 2\lambda - .3$ and $\hat{p} = 2\hat{\lambda} - .3 = 2\left(\dfrac{Y}{n}\right) - .3$; the

estimate is $2\left(\dfrac{20}{80}\right) - .3 = .2$.

b. $E(\hat{p}) = E(2\hat{\lambda} - .3) = 2E(\hat{\lambda}) - .3 = 2\lambda - .3 = p$, as desired.

c. Here $\lambda = .7p + (.3)(.3)$, so $p = \dfrac{10}{7}\lambda - \dfrac{9}{70}$ and $\hat{p} = \dfrac{10}{7}\left(\dfrac{Y}{n}\right) - \dfrac{9}{70}$.

Section 6.2

21.

a. $E(X) = \beta \cdot \Gamma\left(1 + \dfrac{1}{\alpha}\right)$ and $E(X^2) = Var(X) + [E(X)]^2 = \beta^2 \Gamma\left(1 + \dfrac{2}{\alpha}\right)$, so the

moment estimators $\hat{\alpha}$ and $\hat{\beta}$ are the solution to $\overline{X} = \hat{\beta} \cdot \Gamma\left(1 + \dfrac{1}{\hat{\alpha}}\right)$,

$\dfrac{1}{n}\sum X_i^2 = \hat{\beta}^2 \Gamma\left(1 + \dfrac{2}{\hat{\alpha}}\right)$. Thus $\hat{\beta} = \dfrac{\overline{X}}{\Gamma\left(1 + \dfrac{1}{\hat{\alpha}}\right)}$, so once $\hat{\alpha}$ has been determined

$\Gamma\left(1 + \dfrac{1}{\hat{\alpha}}\right)$ is evaluated and $\hat{\beta}$ then computed. Since $\overline{X}^2 = \hat{\beta}^2 \cdot \Gamma^2\left(1 + \dfrac{1}{\hat{\alpha}}\right)$,

$\dfrac{1}{n}\sum \dfrac{X_i^2}{\overline{X}^2} = \dfrac{\Gamma\left(1 + \dfrac{2}{\hat{\alpha}}\right)}{\Gamma^2\left(1 + \dfrac{1}{\hat{\alpha}}\right)}$, so this equation must be solved to obtain $\hat{\alpha}$.

b. From a, $\dfrac{1}{20}\left(\dfrac{16{,}500}{28.0^2}\right) = 1.05 = \dfrac{\Gamma\left(1 + \dfrac{2}{\hat{\alpha}}\right)}{\Gamma^2\left(1 + \dfrac{1}{\hat{\alpha}}\right)}$, so $\dfrac{1}{1.05} = .95 = \dfrac{\Gamma^2\left(1 + \dfrac{1}{\hat{\alpha}}\right)}{\Gamma\left(1 + \dfrac{2}{\hat{\alpha}}\right)}$, and

from the hint, $\dfrac{1}{\hat{\alpha}} = .2 \Rightarrow \hat{\alpha} = 5$. Then $\hat{\beta} = \dfrac{\overline{x}}{\Gamma(1.2)} = \dfrac{28.0}{\Gamma(1.2)}$.

23. For a single sample from a Poisson distribution,

$f(x_1, \ldots, x_n; \lambda) = \dfrac{e^{-\lambda}\lambda^{x_1}}{x_1!} \cdots \dfrac{e^{-\lambda}\lambda^{x_n}}{x_n!} = \dfrac{e^{-n\lambda}\lambda^{\sum x_i}}{x_1! \ldots x_n!}$, so

$\ln[f(x_1, \ldots, x_n; \lambda)] = -n\lambda + \sum x_i \ln(\lambda) - \sum \ln(x_i!)$. Thus

$\dfrac{d}{d\lambda}[\ln[f(x_1, \ldots, x_n; \lambda)]] = -n + \dfrac{\sum x_i}{\lambda} = 0 \Rightarrow \hat{\lambda} = \dfrac{\sum x_i}{n} = \overline{x}$. For our problem,

$f(x_1, \ldots, x_n, y_1 \ldots y_n; \lambda_1, \lambda_2)$ is a product of the x sample likelihood and the y sample

likelihood, implying that $\hat{\lambda}_1 = \overline{x}$, $\hat{\lambda}_2 = \overline{y}$, and (by the invariance principle)

$\widehat{(\lambda_1 - \lambda_2)} = \overline{x} - \overline{y}$.

Chapter 6: Point Estimation

25.

a. $\hat{\mu} = \bar{x} = 384.4; s^2 = 395.16$, so $\frac{1}{n}\sum(x_i - \bar{x})^2 = \hat{\sigma}^2 = \frac{9}{10}(395.16) = 355.64$ and

$\hat{\sigma} = \sqrt{355.64} = 18.86$ (this is not s).

b. The 95th percentile is $\mu + 1.645\sigma$, so the mle of this is (by the invariance principle) $\hat{\mu} + 1.645\hat{\sigma} = 415.42$.

27.

a. $f(x_1,...,x_n;\alpha,\beta) = \dfrac{(x_1 x_2 ...x_n)^{\alpha-1} e^{-\Sigma x_i/\beta}}{\beta^{n\alpha}\Gamma^n(\alpha)}$, so the log likelihood is

$(\alpha-1)\Sigma\ln(x_i) - \dfrac{\Sigma x_i}{\beta} - n\alpha\ln(\beta) - n\ln\Gamma(\alpha)$. Equating both $\dfrac{d}{d\alpha}$ and $\dfrac{d}{d\beta}$ to 0

yields $\Sigma\ln(x_i) - n\ln(\beta) - n\dfrac{d}{d\alpha}\Gamma(\alpha) = 0$ and $\dfrac{\Sigma x_i}{\beta^2} = \dfrac{n\alpha}{\beta} = 0$, a very difficult

system of equations to solve.

b. From the second equation in **a**, $\dfrac{\Sigma x_i}{\beta} = n\alpha \Rightarrow \bar{x} = \alpha\beta = \mu$, so the mle of μ is $\bar{X}$.

29.

a. The joint pdf (aka the likelihood function) is

$$f(x_1,...,x_n;\lambda,\theta) = \begin{cases} \lambda^n e^{-\lambda\Sigma(x_i-\theta)} & x_1 \geq \theta,...,x_n \geq \theta \\ 0 & otherwise \end{cases}$$

But $x_1 \geq \theta,...,x_n \geq \theta$ iff $\min(x_i) \geq \theta$, and $-\lambda\Sigma(x_i-\theta) = -\lambda\Sigma x_i + n\lambda\theta$.

Thus, likelihood $= \begin{cases} \lambda^n \exp(-\lambda\Sigma x_i)\exp(n\lambda\theta) & \min(x_i) \geq \theta \\ 0 & \min(x_i) < \theta \end{cases}$

Consider maximization wrt θ . Because $n\lambda > 0$, increasing θ will increase the likelihood provided that $\min(x_i) \geq \theta$; if we make θ larger than $\min(x_i)$, the likelihood drops to 0. This implies that the mle of θ is $\hat{\theta} = \min(x_i)$. The log likelihood is now $n\ln(\lambda) - \lambda\Sigma(x_i - \hat{\theta})$. Equating the derivative wrt λ to 0 and

solving yields $\hat{\lambda} = \dfrac{n}{\Sigma(x_i - \hat{\theta})} = \dfrac{n}{\Sigma x_i - n\hat{\theta}}$.

b. $\hat{\theta} = \min(x_i) = .64$, and $\Sigma x_i = 55.80$, so $\hat{\lambda} = \dfrac{10}{55.80 - 6.4} = .202$

Supplementary Exercises

31. Substitute $k = \varepsilon/\sigma_Y$ into Chebyshev's inequality to write $P(|Y - \mu_Y| \geq \varepsilon) \leq 1/(\varepsilon/\sigma_Y)^2 = V(Y)/\varepsilon^2$. Since $E(\overline{X}) = \mu$ and $V(\overline{X}) = \sigma^2/n$, we may then write

$P(|\overline{X} - \mu| \geq \varepsilon) \leq \dfrac{\sigma^2/n}{\varepsilon^2}$. As $n \to \infty$, this fraction converges to 0, hence $P(|\overline{X} - \mu| \geq \varepsilon) \to$

0, as desired.

33. Let X_1 = the time until the first birth, X_2 = the elapsed time between the first and second births, and so on. Then

$f(x_1,\ldots,x_n;\lambda) = \lambda e^{-\lambda x_1} \cdot (2\lambda)e^{-2\lambda x_2} \ldots (n\lambda)e^{-n\lambda x_n} = n!\,\lambda^n e^{-\lambda\Sigma kx_k}$. Thus the log

likelihood is $\ln(n!) + n\ln(\lambda) - \lambda\Sigma kx_k$. Taking $\dfrac{d}{d\lambda}$ and equating to 0 yields

$\hat{\lambda} = \dfrac{n}{\sum\limits_{k=1}^{n} kx_k}$. For the given sample, $n = 6$, $x_1 = 25.2$, $x_2 = 41.7 - 25.2 = 16.5$, $x_3 = 9.5$,

$x_4 = 4.3$, $x_5 = 4.0$, $x_6 = 2.3$; so $\sum\limits_{k=1}^{6} kx_k = (1)(25.2) + (2)(16.5) + \ldots + (6)(2.3) = 137.7$

and $\hat{\lambda} = \dfrac{6}{137.7} = .0436$.

35.

$x_i + x_j$	23.5	26.3	28.0	28.2	29.4	29.5	30.6	31.6	33.9	49.3
23.5	23.5	24.9	25.75	25.85	26.45	26.5	27.05	27.55	28.7	36.4
26.3		26.3	27.15	27.25	27.85	27.9	28.45	28.95	30.1	37.8
28.0			28.0	28.1	28.7	28.75	29.3	29.8	30.95	38.65
28.2				28.2	28.8	28.85	29.4	29.9	31.05	38.75
29.4					29.4	29.45	30.0	30.5	30.65	39.35
29.5						29.5	30.05	30.55	31.7	39.4
30.6							30.6	31.1	32.25	39.95
31.6								31.6	32.75	40.45
33.9									33.9	41.6
49.3										49.3

There are 55 averages, so the median is the 28th in order of increasing magnitude. Therefore, $\hat{\mu} = 29.5$

37. Let $c = \dfrac{\Gamma\left(\frac{n-1}{2}\right)}{\Gamma\left(\frac{n}{2}\right) \cdot \sqrt{\frac{2}{n-1}}}$. Then $E(cS) = cE(S)$, and c cancels with the two Γ factors and the square root in $E(S)$, leaving just σ. When $n = 20$, $c = \dfrac{\Gamma(9.5)}{\Gamma(10) \cdot \sqrt{\frac{2}{19}}}$. $\Gamma(10) = 9!$ and $\Gamma(9.5) = (8.5)(7.5)...(1.5)(.5)\Gamma(.5)$, but $\Gamma(.5) = \sqrt{\pi}$. Straightforward calculation gives $c = 1.0132$.

CHAPTER 7

Section 7.1

1.

$\bar{x}(c) = 0.9$

   a. $z_{\alpha/2} = 2.81$ implies that $\alpha/2 = 1 - \Phi(2.81) = .0025$, so $\alpha = .005$ and the confidence level is $100(1 - \alpha)\% = 99.5\%$.

   b. $z_{\alpha/2} = 1.44$ for $\alpha = 2[1 - \Phi(1.44)] = .15$, and $100(1 - \alpha)\% = 85\%$.

   c.   99.7% implies that $\alpha = .003$, $\alpha/2 = .0015$, and $z_{.0015} = 2.96$. (Look for cumulative area .9985 in the main body of the z table.)

   d.   75% implies $\alpha = .25$, $\alpha/2 = .125$, and $z_{.125} = 1.15$.

3.

   a.   A 90% confidence interval will be narrower. The z critical value for a 90% confidence level is 1.645, smaller than the z of 1.96 for the 95% confidence level, thus producing a narrower interval.

   b.   Not a correct statement. Once and interval has been created from a sample, the mean μ is either enclosed by it, or not. We have 95% confidence in the general procedure, under repeated and independent sampling.

   c.   Not a correct statement. The interval is an estimate for the population mean, not a boundary for population values.

   d.   Not a correct statement. In theory, if the process were repeated an infinite number of times, 95% of the intervals would contain the population mean μ. We *expect* 95 out of 100 intervals will contain μ, but we don't know this to be true.

5.

a. $4.85 \pm \dfrac{(1.96)(.75)}{\sqrt{20}} = 4.85 \pm .33 = (4.52, 5.18)$.

b. $z_{\alpha/2} = z_{.02/2} = z_{.01} = 2.33$, so the interval is $4.56 \pm \dfrac{(2.33)(.75)}{\sqrt{16}} = (4.12, 5.00)$.

c. $n \geq \left[\dfrac{2(1.96)(.75)}{.40} \right]^2 = 54.02$, so $n = 55$.

d. $w = 2(.2) = .4$, so $n \geq \left[\dfrac{2(2.58)(.75)}{.4} \right]^2 = 93.61$, so $n = 94$.

7. If $L = 2z_{\alpha/2} \dfrac{\sigma}{\sqrt{n}}$ and we increase the sample size by a factor of 4, the new length is

$L' = 2z_{\alpha/2} \dfrac{\sigma}{\sqrt{4n}} = \left[2z_{\alpha/2} \dfrac{\sigma}{\sqrt{n}} \right] \left(\dfrac{1}{2} \right) = \dfrac{L}{2}$. Thus halving the length requires n to be

increased 4-fold. If $n' = 25n$, then $L' = \dfrac{L}{5}$, so the length is decreased by a factor of 5.

9.

a. $\left(\bar{x} - 1.645 \dfrac{\sigma}{\sqrt{n}}, \infty \right)$. From 5a, $\bar{x} = 4.85$, $\sigma = .75$, $n = 20$;

$4.85 - 1.645 \dfrac{.75}{\sqrt{20}} = 4.5741$, so the interval is $(4.5741, \infty)$.

b. $\left(\bar{x} - z_{\alpha} \dfrac{\sigma}{\sqrt{n}}, \infty \right)$

c. $\left(-\infty, \bar{x} + z_{\alpha} \dfrac{\sigma}{\sqrt{n}} \right)$; From 4a, $\bar{x} = 58.3$, $\sigma = 3.0$, $n = 25$;

$58.3 + 2.33 \dfrac{3}{\sqrt{25}} = (-\infty, 59.70)$

11.   Y is a binomial r.v. with $n = 1000$ and $p = .95$, so $E(Y) = np = 950$, the expected
        number of intervals that capture μ, and $\sigma_Y = \sqrt{npq} = 6.892$. Using the normal
        approximation to the binomial distribution, $P(940 \leq Y \leq 960) = P(939.5 \leq Y_{normal} \leq$
        $960.5) = P(-1.52 \leq Z \leq 1.52) = .9357 - .0643 = .8714$.

Section 7.2

13.

a. $\bar{x} \pm z_{.025} \dfrac{s}{\sqrt{n}} = 654.16 \pm 1.96 \dfrac{164.43}{\sqrt{50}} = (608.58, 699.74)$. We are 95% confident
        that the true average CO_2 level in this population of homes with gas cooking
        appliances is between 608.58ppm and 699.74ppm

b. $w = 50 = \dfrac{2(1.96)(175)}{\sqrt{n}} \Rightarrow n = \left(\dfrac{2(1.96)(175)}{50} \right)^2 = 188.24 \uparrow 189$

15.

a. $z_\alpha = .84$, and $\Phi(.84) = .7995 \approx .80$, so the confidence level is 80%.

b. $z_\alpha = 2.05$, and $\Phi(2.05) = .9798 \approx .98$, so the confidence level is 98%.

c. $z_\alpha = .67$, and $\Phi(.67) = .7486 \approx .75$, so the confidence level is 75%.

17. $\bar{x} - z_{.01} \dfrac{s}{\sqrt{n}} = 135.39 - 2.33 \dfrac{4.59}{\sqrt{153}} = 135.39 - .865 = 134.53$. We are 99% confident
        that the true average ultimate tensile strength is greater than 134.53.

19. $\hat{p} = \dfrac{201}{356} = .5646$; We calculate a 95% confidence interval for the proportion of all dies that pass the probe:

$$\frac{.5646 + \dfrac{(1.96)^2}{2(356)} \pm 1.96\sqrt{\dfrac{(.5646)(.4354)}{356} + \dfrac{(1.96)^2}{4(356)^2}}}{1 + \dfrac{(1.96)^2}{356}} = \frac{.5700 \pm .0518}{1.01079} = (.513, .615)$$

21. $\hat{p} = \dfrac{133}{539} = .2468$; the 95% lower confidence bound is:

$$\frac{.2468 + \dfrac{(1.645)^2}{2(539)} - 1.645\sqrt{\dfrac{(.2468)(.7532)}{539} + \dfrac{(1.645)^2}{4(539)^2}}}{1 + \dfrac{(1.645)^2}{539}} = \frac{.2493 - .0307}{1.005} = .218$$

23.

a. $\hat{p} = \dfrac{24}{37} = .6486$; The 99% confidence interval for p is

$$\frac{.6486 + \dfrac{(2.58)^2}{2(37)} \pm 2.58\sqrt{\dfrac{(.6486)(.3514)}{37} + \dfrac{(2.58)^2}{4(37)^2}}}{1 + \dfrac{(2.58)^2}{37}} = \frac{.7386 \pm .2216}{1.1799} = (.438, .814)$$

b. $n = \dfrac{2(2.58)^2(.25) - (2.58)^2(.01) + \sqrt{4(2.58)^4(.25)(.25 - .01) + .01(2.58)^4}}{.01}$

$= \dfrac{3.261636 + 3.3282}{.01} \approx 659$

25.

a. $n = \dfrac{2(1.96)^2(.25) - (1.96)^2(.01) + \sqrt{4(1.96)^4(.25)(.25 - .01) + .01(1.96)^4}}{.01} \approx 381$

b. $n = \dfrac{2(1.96)^2\left(\frac{1}{3} \cdot \frac{2}{3}\right) - (1.96)^2(.01) + \sqrt{4(1.96)^4\left(\frac{1}{3} \cdot \frac{2}{3}\right)\left(\frac{1}{3} \cdot \frac{2}{3} - .01\right) + .01(1.96)^4}}{.01} \approx 339$

27.   Note that the midpoint of the new interval is $\dfrac{x + \frac{z^2}{2}}{n + z^2}$, which is roughly $\dfrac{x+2}{n+4}$ with a confidence level of 95% and approximating $1.96 \approx 2$. The variance of this quantity is $\dfrac{np(1-p)}{\left(n + z^2\right)^2}$, or roughly $\dfrac{p(1-p)}{n+4}$. Now replacing p with $\dfrac{x+2}{n+4}$, we have

$$\left(\frac{x+2}{n+4}\right) \pm z_{\alpha/2} \sqrt{\frac{\left(\dfrac{x+2}{n+4}\right)\left(1 - \dfrac{x+2}{n+4}\right)}{n+4}} \ .$$

For clarity, let $x^* = x + 2$ and $n^* = n + 4$, then $\hat{p}^* = \dfrac{x^*}{n^*}$ and the formula reduces to

$$\hat{p}^* \pm z_{\alpha/2} \sqrt{\frac{\hat{p}^* \hat{q}^*}{n^*}} \ ,$$ the desired conclusion. For further discussion, see the Agresti article.

Section 7.3

29.

   a. $t_{.025,10} = 2.228$ **d.** $t_{.005,50} = 2.678$

   b. $t_{.025,20} = 2.086$ **e.** $t_{.01,25} = 2.485$

   c. $t_{.005,20} = 2.845$ **f.** $-t_{.025,5} = -2.571$

31.

   a. $t_{05,10} = 1.812$ **d.** $t_{.01,4} = 3.747$

   b. $t_{.05,15} = 1.753$ **e.** $\approx t_{.025,24} = 2.064$

   c. $t_{.01,15} = 2.602$ **f.** $t_{.01,37} \approx 2.429$

33.

a. The boxplot indicates a very slight positive skew, with no outliers. The data appears to center near 438.

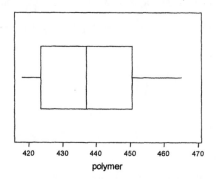

polymer

b. Based on a normal probability plot, it is reasonable to assume the sample observations came from a normal distribution.

c. With d.f. $= n - 1 = 16$, the critical value for a 95% C.I. is $t_{.025,16} = 2.120$, and the interval is $438.29 \pm (2.120)\left(\dfrac{15.14}{\sqrt{17}}\right) = 438.29 \pm 7.785 = (430.51, 446.08)$. Since 440 is within the interval, 440 is a plausible value for the true mean. 450, however, is not, since it lies outside the interval.

35. $n = 15$, $\bar{x} = 25.0$, $s = 3.5$; $t_{.025,15} = 2.131$

a. A 95% C.I. for the mean: $25.0 \pm 2.131\dfrac{3.5}{\sqrt{15}} = (23.1, 26.9)$

b. A 95% Prediction Interval: $25.0 \pm 2.131(3.5)\sqrt{1 + \dfrac{1}{15}} = (17.3, 32.7)$. The prediction interval is about 4 times wider than the confidence interval.

37.

a. A 95% C.I. : $.9255 \pm 2.093(.0181) = .9255 \pm .0379 \Rightarrow (.8876, .9634)$

b. A 95% P.I. : $.9255 \pm 2.093(.0809)\sqrt{1 + \frac{1}{20}} = .9255 \pm .1735 \Rightarrow (.7520, 1.0990)$

c. A tolerance interval is requested, with k = 99, confidence level 95%, and n = 20. The tolerance critical value, from Table A.6, is 3.615. The interval is $.9255 \pm 3.615(.0809) \Rightarrow (.6330, 1.2180)$.

39.

a.

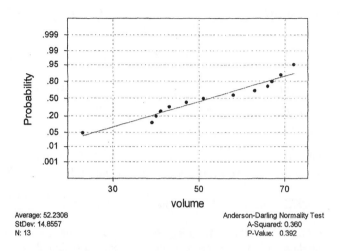

Normal Probability Plot

Average: 52.2308
StDev: 14.8557
N: 13

Anderson-Darling Normality Test
A-Squared: 0.360
P-Value: 0.392

Based on the above plot, generated by Minitab, it is plausible that the population distribution is normal.

b. We require a tolerance interval. From table A6, with 95% confidence, k = 95, and n=13, the table critical value = 3.081.

$$\bar{x} \pm (tcv)s = 52.231 \pm 3.081(14.856) = 52.231 \pm 45.771 \Rightarrow (6.460, 98.002)$$

c. A prediction interval, with $t_{.025,12} = 2.179$:

$$52.231 \pm 2.179(14.856)\sqrt{1 + \tfrac{1}{13}} = 52.231 \pm 33.593 \Rightarrow (18.638, 85.824)$$

41. The 20 d.f. row of Table A.5 shows that 1.725 captures upper tail area .05 and 1.325 captures upper tail area .10 The confidence level for each interval is 100(central area)%. For the first interval, central area = 1 − sum of tail areas = 1 − (.25 + .05) = .70, and for the second and third intervals the central areas are 1 − (.20 + .10) = .70 and 1 − (.15 + .15) = 70. Thus each interval has confidence level 70%. The width of the first interval is $\dfrac{s(.687 + 1.725)}{\sqrt{n}} = \dfrac{2.412s}{\sqrt{n}}$, whereas the widths of the second and third intervals are 2.185 and 2.128 standard errors respectively. The third interval, with symmetrically placed critical values, is the shortest, so it should be used. This will always be true for a t interval.

Section 7.4

43.

    a. $\chi^2_{.05,10} = 18.307$ **b.** $\chi^2_{.95,10} = 3.940$

    c. Since $10.987 = \chi^2_{.975,22}$ and $36.78 = \chi^2_{.025,22}$, $P\left(\chi^2_{.975,22} \le \chi^2 \le \chi^2_{.025,22}\right) = .95$.

    d. Since $14.611 = \chi^2_{.95,25}$ and $37.652 = \chi^2_{.05,25}$, $P(\chi^2 < 14.611 \text{ or } \chi^2 > 37.652) =$
        $1 - P(\chi^2 > 14.611) + P(\chi^2 > 37.652) = (1 - .95) + .05 = .10$.

45. $n = 22$ implies that d.f. $= n - 1 = 21$, so the .995 and .005 columns of Table A.7 give
the necessary chi-squared critical values as 8.033 and 41.399. $\Sigma x_i = 1701.3$ and

$\Sigma x_i^2 = 132,097.35$, so $s^2 = 25.368$. The interval for σ^2 is $\left(\dfrac{21(25.368)}{41.399}, \dfrac{21(25.368)}{8.033} \right)$

$= (12.868, 66.317)$ and the CI for σ is $(3.6, 8.1)$. Validity of this interval requires that
fracture toughness be at least approximately normally distributed.

Supplementary Exercises

47.

    a. $n = 48$, $\bar{x} = 8.079$, $s^2 = 23.7017$, and $s = 4.868$.
        A 95% C.I. for $\mu =$ the true average strength is

$$\bar{x} \pm 1.96 \frac{s}{\sqrt{n}} = 8.079 \pm 1.96 \frac{4.868}{\sqrt{48}} = 8.079 \pm 1.377 = (6.702, 9.456)$$

    b. $\hat{p} = \dfrac{13}{48} = .2708$. A 95% C.I. is

$$\frac{.2708 + \dfrac{1.96^2}{2(48)} \pm 1.96 \sqrt{\dfrac{(.2708)(.7292)}{48} + \dfrac{1.96^2}{4(48)^2}}}{1 + \dfrac{1.96^2}{48}} = \frac{.3108 \pm .1319}{1.0800} = (.166, .410)$$

49.

a. There appears to be a slight positive skew in the middle half of the sample, but the lower whisker is much longer than the upper whisker. The extent of variability is rather substantial, although there are no outliers.

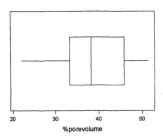

b. The pattern of points in a normal probability plot is reasonably linear, so, yes, normality is plausible.

c. $n = 18$, $\bar{x} = 38.66$, $s = 8.473$, and $t_{.01,17} = 2.567$. The 98% confidence interval is

$$38.66 \pm 2.567 \frac{8.473}{\sqrt{18}} = 38.66 \pm 5.13 = (33.53, 43.79).$$

51.

a. $\hat{p} = \dfrac{136}{200} = .680 \Rightarrow$ a 90% C.I. is

$$\frac{.680 + \dfrac{1.645^2}{2(200)} \pm 1.645 \sqrt{\dfrac{(.680)(.320)}{200} + \dfrac{1.645^2}{4(200)^2}}}{1 + \dfrac{1.645^2}{200}} = \frac{.6868 \pm .0547}{1.01353} = (.624, .732)$$

b. $n = \dfrac{2(1.645)^2(.25) - (1.645)^2(.05)^2 + \sqrt{4(1.645)^4(.25)(.25 - .0025) + .05^2(1.645)^4}}{.0025}$

$$= \frac{1.3462 + 1.3530}{.0025} = 1079.7 \uparrow 1080$$

c. No, it gives a 95% upper bound.

53. With $\hat{\theta} = \frac{1}{3}(\overline{X}_1 + \overline{X}_2 + \overline{X}_3) - \overline{X}_4$, $\sigma_{\hat{\theta}}^2 = \frac{1}{9}Var(\overline{X}_1 + \overline{X}_2 + \overline{X}_3) + Var(\overline{X}_4) =$

$\frac{1}{9}\left(\frac{\sigma_1^2}{n_1} + \frac{\sigma_2^2}{n_2} + \frac{\sigma_3^2}{n_3}\right) + \frac{\sigma_4^2}{n_4}$; $\hat{\sigma}_{\hat{\theta}}$ is obtained by replacing each σ_i^2 by s_i^2 and taking

the square root. The large-sample interval for θ is then

$\frac{1}{3}(\overline{x}_1 + \overline{x}_2 + \overline{x}_3) - \overline{x}_4 \pm z_{\alpha/2}\sqrt{\frac{1}{9}\left(\frac{s_1^2}{n_1} + \frac{s_2^2}{n_2} + \frac{s_3^2}{n_3}\right) + \frac{s_4^2}{n_4}}$. For the given data, $\hat{\theta} = -.50$

and $\hat{\sigma}_{\hat{\theta}} = .1718$, so the interval is $-.50 \pm 1.96(.1718) = (-.84, -.16)$.

55. The specified condition is that the interval be length .2, so $n \geq \left[\frac{2(1.96)(.8)}{.2}\right]^2 = 245.86$,

and $n = 246$ should be used.

57. Proceeding as in Example 7.5 with T_r replacing ΣX_i, the C.I. for $\frac{1}{\lambda}$ is

$\left(\frac{2t_r}{\chi_{1-\alpha/2, 2r}^2}, \frac{2t_r}{\chi_{\alpha/2, 2r}^2}\right)$ where $t_r = y_1 + ... + y_r + (n-r)y_r$.

In Example 6.8, $n = 20$, $r = 10$, and $t_r = 1115$. With d.f. = 20, the necessary critical values are 9.591 and 34.170, giving the interval (65.3, 232.5). This is obviously an extremely wide interval. The censored experiment provides less information about $\frac{1}{\lambda}$ than would an uncensored experiment with $n = 20$.

59.

a. $\int_{(\alpha/2)^{1/n}}^{(1-\alpha/2)^{1/n}} nu^{n-1} du = u^n \Big|_{(\alpha/2)^{1/n}}^{(1-\alpha/2)^{1/n}} = 1 - \dfrac{\alpha}{2} - \dfrac{\alpha}{2} = 1 - \alpha$. From the probability

statement, $\dfrac{(\alpha/2)^{\frac{1}{n}}}{\max(X_i)} \le \dfrac{1}{\theta} \le \dfrac{(1-\alpha/2)^{\frac{1}{n}}}{\max(X_i)}$ with probability $1-\alpha$, so taking the

reciprocal of each endpoint and interchanging gives the C.I.

$\left(\dfrac{\max(X_i)}{(1-\alpha/2)^{\frac{1}{n}}}, \dfrac{\max(X_i)}{(\alpha/2)^{\frac{1}{n}}} \right)$ for θ.

b. $\alpha^{\frac{1}{n}} \le \dfrac{\max(X_i)}{\theta} \le 1$ with probability $1-\alpha$, so $1 \le \dfrac{\theta}{\max(X_i)} \le \dfrac{1}{\alpha^{\frac{1}{n}}}$ with

probability $1-\alpha$, which yields the interval $\left(\max(X_i), \dfrac{\max(X_i)}{\alpha^{\frac{1}{n}}} \right)$.

c. It is easily verified that the interval of **b** is shorter – draw a graph of $f_U(u)$ and verify that the shortest interval which captures area $1-\alpha$ under the curve is the rightmost such interval, which leads to the C.I. of **b**. With $\alpha = .05$, $n = 5$, max(x_i)=4.2; this yields (4.2, 7.65).

61. $\tilde{x} = 76.2$, the lower and upper fourths are 73.5 and 79.7, respectively, and $f_s = 6.2$.

The robust interval is $76.2 \pm (1.93)\left(\dfrac{6.2}{\sqrt{22}} \right) = 76.2 \pm 2.6 = (73.6, 78.8)$.

$\bar{x} = 77.33$, $s = 5.037$, and $t_{.025,21} = 2.080$, so the t interval is

$77.33 \pm (2.080)\left(\dfrac{5.037}{\sqrt{22}} \right) = 77.33 \pm 2.23 = (75.1, 79.6)$. The t interval is centered at $\bar{x}$,

which is pulled out to the right of $\tilde{x}$ by the single mild outlier 93.7; the interval widths are comparable.

CHAPTER 8

Section 8.1

1.

    a.   Yes. It is an assertion about the value of a parameter.

    b.   No. The sample median $\tilde{x}$ is not a parameter.

    c.   No. The sample standard deviation s is not a parameter.

    d.   Yes. The assertion is that the standard deviation of population #2 exceeds that of population #1.

    e.   No. $\overline{X}$ and $\overline{Y}$ are statistics rather than parameters, so they cannot appear in a hypothesis.

    f.   Yes. H is an assertion about the value of a parameter.

3.   In this formulation, H_0 states the welds do not conform to specification. This assertion will not be rejected unless there is strong evidence to the contrary. Thus the burden of proof is on those who wish to assert that the specification is satisfied. Using $H_a: \mu < 100$ results in the welds being believed in conformance unless proved otherwise, so the burden of proof is on the non–conformance claim.

5.   Let σ denote the population standard deviation. Then the appropriate hypotheses are $H_0: \sigma = .05$ v. $H_a: \sigma < .05$. With this formulation, the burden of proof is on the data to show that the requirement has been met (the sheaths will not be used unless H_0 can be rejected in favor of H_a). Type I error: Conclude that the standard deviation is $< .05$ mm when it is really equal to .05 mm. Type II error: Conclude that the standard deviation is .05 mm when it is really $< .05$.

7.   A type I error here involves saying that the plant is not in compliance when in fact it is. A type II error occurs when we conclude that the plant is in compliance when in fact it isn't. Reasonable people may disagree as to which of the two errors is more serious. If in your judgement it is the type II error, then the reformulation $H_0: \mu = 150$ v. $H_a: \mu < 150$ makes the type I error more serious.

9.

a. R_1 is most appropriate, because x either too large or too small contradicts p = .5 and supports p ≠ .5.

b. A type I error consists of judging one of the two companies favored over the other when in fact there is a 50–50 split in the population. A type II error involves judging the split to be 50–50 when it is not.

c. X has a binomial distribution with n = 25 and p = 0.5. α = P(type I error) = $P(X \le 7 \, or \, X \ge 18$ when X ~ Bin(25, .5)) = B(7; 25,.5) + 1 – B(17; 25,.5) = .044

d. $\beta(.4) = P(8 \le X \le 17$ when p = .4) = B(17; 25,.5) – B(7, 25,.4) = 0.845, and $\beta(.6) = 0.845$ also. $\beta(.3) = B(17;25,.3) – B(7;25,.3) = .488 = \beta(.7)$

e. x = 6 is in the rejection region R_1, so H_0 is rejected in favor of H_a.

11.

a. H_0: $\mu = 10$ v. H_a: $\mu \ne 10$

b. α = Prejecting H_0 when H_0 is true) = $P(\bar{x} \ge 10.1032 \, or \le 9.8968$ when $\mu = 10..$

Since $\bar{x}$ is normally distributed with standard deviation $\dfrac{\sigma}{\sqrt{n}} = \dfrac{.2}{5} = .04$,

$\alpha = P(z \ge 2.58 \, or \le -2.58) = .005 + .005 = .01$

c. When $\mu = 10.1$, $E(\bar{x}) = 10.1$ so $\beta(10.1) = P(9.8968 < \bar{x} < 10.1032$ when $\mu = 10.1)$ = $P(-5.08 < z < .08) = .5319$. Similarly, $\beta(9.8) = P(2.42 < z < 7.58) = .0078$

d. $c = \pm 2.58$

e. Now $\dfrac{\sigma}{\sqrt{n}} = \dfrac{.2}{3.162} = .0632$. Thus 10.1032 is replaced by c, where $\dfrac{c - 10}{.0632} = 1.96$ and so c = 10.124. Similarly, 9.8968 is replaced by 9.876.

f. $\bar{x} = 10.020$. Since $\bar{x}$ is neither ≥ 10.124 nor ≤ 9.876, it is not in the rejection region. H_0 is not rejected; it is still plausible that $\mu = 10$

g. $\bar{x} \ge 10.1032$ or ≤ 9.8968 iff $z \ge 2.58$ or ≤ -2.58.

13.

a. $P(\bar{x} \geq \mu_o + 2.33 \frac{\sigma}{\sqrt{n}}$ when $\mu = \mu_o) = P\left(Z \geq \frac{\left(\mu_o + 2.33\sigma / \sqrt{n} - \mu_o\right)}{\sigma / \sqrt{n}}\right)$

$= P(z \geq 2.33) = .01$, where Z is a standard normal r.v.

b. P(rejecting H_0 when $\mu = 99$) $= P(\bar{x} \geq 102.33$ when $\mu = 99)$ $= P\left(z \geq \frac{102 - 99}{1}\right)$

$= P(z \geq 3.33) = .0004$. Similarly, $\alpha(98) = P(\bar{x} \geq 102.33$ when $\mu = 98) = P(Z \geq$
$4.33 = 0$. In general, we have P(type I error) $< .01$ when this probability is
calculated for a value of μ less than 100. The boundary value $\mu = 100$ yields the
largest α.

Section 8.2

15.

a. $\alpha = P(z \geq 1.88$ when z has a standard normal distribution) $= 1 - \Phi(1.88) = .0301$

b. $\alpha = P(z \leq -2.75$ when $z \sim$ N(0, 1) $= \Phi(-2.75) = .003$

c. $\alpha = \Phi(-2.88) + (1 - \Phi(2.88)) = .004$

17.

a. $z = \dfrac{30,960 - 30,000}{1500/\sqrt{16}} = 2.56 > 2.33$ so reject H_0.

b. $\beta(30,500): \Phi\left(2.33 + \dfrac{30,000 - 30,500}{1500/\sqrt{16}}\right) = \Phi(1.00) = .8413$

c. $\beta(30,500) = .05 : n = \left[\dfrac{1500(2.33 + 1.645)}{30,000 - 30,500}\right]^2 = 142.2$, so use n = 143

d. $\alpha = 1 - \Phi(2.56) = .0052$

19.

a. Reject H_0 if either $z \geq 2.58$ or $z \leq -2.58$; $\dfrac{\sigma}{\sqrt{n}} = 0.3$, so

$$z = \frac{94.32 - 95}{0.3} = -2.27.$$ Since -2.27 is not < -2.58, don't reject H_0.

b. $\beta(94) = \Phi\left(2.58 + \dfrac{1}{0.3}\right) - \Phi\left(-2.58 + \dfrac{1}{0.3}\right) = \Phi(5.91) - \Phi(.75) = .2266$

c. $n = \left[\dfrac{1.20(2.58 + 1.28)}{95 - 94}\right]^2 = 21.46$, so use $n = 22$.

21. With H_0: $\mu = .5$ and H_a: $\mu \neq .5$, we reject H_0 if $t > t_{\alpha/2, n-1}$ or $t < -t_{\alpha/2, n-1}$

a. $1.6 < t_{.025, 12} = 2.179$, so don't reject H_0

b. $-1.6 > -t_{.025, 12} = -2.179$, so don't reject H_0

c. $-2.6 > -t_{.005, 24} = -2.797$, so don't reject H_0

d. $-3.9 <$ the negative of all t values at df = 24, so we reject H_0 in favor of H_a

23. H_0: $\mu = 360$ v.. H_a: $\mu > 360$; $t = \dfrac{\bar{x} - 360}{s / \sqrt{n}}$; reject H_0 if $t > t_{.05, 25} = 1.708$;

$$t = \frac{370.69 - 360}{24.36 / \sqrt{26}} = 2.24 > 1.708.$$ Thus H_0 should be rejected. There appears to be a contradiction of the prior belief.

25.

a. H_0: $\mu = 5.5$ v. H_a: $\mu \neq 5.5$; for a level .01 test, (not specified in the problem description), reject H_0 if either $z \geq 2.58$ or $z \leq -2.58$. Since

$$z = \frac{5.25 - 5.5}{.075} = -3.33 \leq -2.58,$$ reject H_0.

b. $1 - \beta(5.6) = 1 - \Phi\left(2.58 + \dfrac{(-.1)}{.075}\right) + \Phi\left(-2.58 + \dfrac{(-.1)}{.075}\right)$

$= 1 - \Phi(1.25) + \Phi(-3.91) = .105$

c. $n = \left[\dfrac{.3(2.58 + 2.33)}{-.1}\right]^2 = 216.97$, so use $n = 217$.

Chapter 8: Tests of Hypotheses Based on a Single Sample

27.

a. Using software, $\bar{x} = 0.75$, $\tilde{x} = 0.64$, $s = .3025$, $f_s = 0.48$. These summary statistics, as well as a box plot (not shown) indicate substantial positive skewness, but no outliers.

b. No, it is not plausible from the results in part **a** that the variable ALD is normal. However, since $n = 49$, normality is not required for the use of z inference procedures.

c. We wish to test $H_0: \mu \geq 1.0$ versus $H_a: \mu < 1.0$. The test statistic is
$$z = \frac{0.75 - 1.0}{.3025 / \sqrt{49}} = -5.79;$$ at any reasonable significance level, we reject the null hypothesis. Yes, the data provides strong evidence that the true average ALD is less than 1.0.

d. $\bar{x} + z_{.05} \dfrac{s}{\sqrt{n}} = 0.75 + 1.645 \dfrac{.3025}{\sqrt{49}} = 0.821$

29.

a. For n = 8, n – 1 = 7, and $t_{.05,7} = 1.895$, so H_0 is rejected at level .05 if $t \geq 1.895$. Since $\dfrac{s}{\sqrt{n}} = \dfrac{1.25}{\sqrt{8}} = .442$, $t = \dfrac{3.72 - 3.50}{.442} = .498$; this does not exceed 1.895, so H_0 is not rejected.

b. $d = \dfrac{|\mu_o - \mu|}{\sigma} = \dfrac{|3.50 - 4.00|}{1.25} = .40$, and df = 7, so from table A.17, $\beta(.40) \approx .73$

31. The hypotheses of interest are $H_0: \mu = 7$ v. $H_a: \mu < 7$, so a lower–tailed test is appropriate; H_0 should be rejected if $t \leq -t_{.1,8} = -1.397$. $t = \dfrac{6.32 - 7}{1.65 / \sqrt{9}} = -1.24$. Because –1.24 is not ≤ -1.397, H_0 is not rejected (contradicted) at level .01.

33. $\beta(\mu_o - \Delta) = \Phi\left(z_{\alpha/2} + \Delta\sqrt{n}/\sigma\right) - \Phi\left(-z_{\alpha/2} + \Delta\sqrt{n}/\sigma\right)$
$= 1 - \Phi\left(-z_{\alpha/2} - \Delta\sqrt{n}/\sigma\right) - [1 - \Phi\left(z_{\alpha/2} - \Delta\sqrt{n}/\sigma\right)] =$
$\Phi\left(z_{\alpha/2} - \Delta\sqrt{n}/\sigma\right) - \Phi\left(-z_{\alpha/2} - \Delta\sqrt{n}/\sigma\right) = \beta(\mu_o + \Delta)$

Section 8.3

35.

1   Parameter of interest: p = true proportion of cars in this particular county passing emissions testing on the first try.

2   H_0: p = .70

3   H_a: p ≠ .70

4    $z = \dfrac{\hat{p} - p_o}{\sqrt{p_o(1 - p_o)/n}} = \dfrac{\hat{p} - .70}{\sqrt{.70(.30)/n}}$

5   either z ≥ 1.96 or z ≤ −1.96

6    $z = \dfrac{124/200 - .70}{\sqrt{.70(.30)/200}} = -2.469$

7   Reject H_0. The data indicates that the proportion of cars passing the first time on emission testing or this county differs from the proportion of cars passing statewide.

37.

1   p = true proportion of all donors with type A blood

2   H_0: p = .40

3   H_a: p ≠ .40

4    $z = \dfrac{\hat{p} - p_o}{\sqrt{p_o(1 - p_o)/n}} = \dfrac{\hat{p} - .40}{\sqrt{.40(.60)/n}}$

5   Reject H_0 if z ≥ 2.58 or z ≤ −2.58

6    $z = \dfrac{82/150 - .40}{\sqrt{.40(.60)/150}} = \dfrac{.147}{.04} = 3.667$

7   Reject H_0. The data does suggest that the percentage of the population with type A blood differs from 40% (at the .01 significance level). Since the z critical value for a significance level of .05 is less than that of .01, the conclusion would not change.

39.

a. We wish to test $H_0: p = .02$ v. $H_a: p < .02$; only if H_0 can be rejected will the inventory be postponed. The lower-tailed test rejects H_0 if $z \le -1.645$. With $\hat{p} = \dfrac{15}{1000} = .015$, $z = -1.01$, which is not ≤ -1.645. Thus, H_0 cannot be rejected, so the inventory should be carried out.

b. $\beta(.01) = 1 - \Phi\left[\dfrac{.02 - .01 - 1.645\sqrt{.02(.98)/1000}}{\sqrt{.01(.99)/1000}}\right] = 1 - \Phi(0.86) = .1949$

c. $\beta(.05) = 1 - \Phi\left[\dfrac{.02 - .05 - 1.645\sqrt{.02(.98)/1000}}{\sqrt{.05(.95)/1000}}\right] = 1 - \Phi(-5.41) \approx 1$, so the chance the inventory will be *postoned* is P(reject H_0 when $p = .05$) $= 1 - \beta(.05) = 0$. It is highly unlikely that H_0 will be rejected, and the inventory will almost surely be carried out.

41.

a. $p =$ true proportion of current customers who qualify. $H_0: p = .05$ v. $H_a: p \ne .05$, $z = \dfrac{\hat{p} - .05}{\sqrt{.05(.95)/n}}$, reject H_0 if $z \ge 2.58$ or $z \le -2.58$. $\hat{p} = .08$, so $z = \dfrac{.03}{.00975} = 3.07 \ge 2.58$, so H_0 is rejected. The company's premise is not correct.

b. $\beta(.10) = \Phi\left[\dfrac{.05 - .10 + 2.58\sqrt{.05(.95)/500}}{\sqrt{.10(.90)/500}}\right] - \Phi\left[\dfrac{.05 - .10 - 2.58\sqrt{.05(.95)/500}}{\sqrt{.10(.90)/500}}\right]$

$\approx \Phi(-1.85) - 0 = .0332$

43. The hypotheses are $H_0: p = .10$ v.. $H_a: p > .10$, so R has the form $\{c, \ldots, n\}$. The values $n = 10$, $c = 3$ (i.e. $R = \{3, 4, \ldots, 10\}$) yield $\alpha = 1 - B(2; 10, .1) = .07$ while no larger R has $\alpha \le .10$; however $\beta(.3) = B(2; 10, .3) = .383$. For $n = 20$, $c = 5$ yields $\alpha = 1 - B(4; 20, .1) = .043$, but again $\beta(.3) = B(4; 20, .3) = .238$. For $n = 25$, $c = 5$ yields $\alpha = 1 - B(4; 25, .1) = .098$ while $\beta(.7) = B(4; 25, .3) = .090 \le .10$, so $n = 25$ should be used. The rejection region is $R = \{5, \ldots, 25\}$, $\alpha = .098$, and $\beta(.7) = .090$.

Section 8.4

45. Using $\alpha = .05$, H_0 should be rejected whenever P-value $< .05$.
  a. P-value $= .001 < .05$, so reject H_0

  b. $.021 < .05$, so reject H_0.

  c. $.078$ is not $< .05$, so don't reject H_0.

  d. $.047 < .05$, so reject H_0 (a close call).

  e. $.148 > .05$, so H_0 can't be rejected at level .05.

47. In each case the p-value $= P(Z > z) = 1 - \Phi(z)$
  a. .0778

  b. .1841

  c. .0250

  d. .0066

  e. .5438

49. Use Table A.8.
  a. $P(t > 2.0)$ at $8df = .040$

  b. $P(t < -2.4)$ at $11df = .018$

  c. $2P(t < -1.6)$ at $15df = 2(.065) = .130$

  d. by symmetry, $P(t > -.4) = 1 - P(t > .4)$ at $19df = 1 - .347 = .653$

  e. $P(t > 5.0)$ at $5df < .005$

  f. $2P(t < -4.8)$ at $40df < 2(.000) = .000$ to three decimal places

51. The p-value is greater than the level of significance $\alpha = .01$, therefore fail to reject H_0 that $\mu = 5.63$. The data does not indicate a statistically significant difference in average serum receptor concentration between pregnant women and all other women.

53. Here we might be concerned with departures above as well as below the specified weight of 5.0, so the relevant hypotheses are $H_0: \mu = 5.0$ v. $H_a: \mu \neq 5.0$. Since $\frac{s}{\sqrt{n}} = .035$, $z = \frac{-.13}{.035} = -3.71$. Because 3.71 is "off" the z-table, P-value $< 2(.0002)$ $= .0004$, so H_0 should be rejected.

55. p = proportion of all physicians that know the generic name for methadone. $H_0: p = .50$ v. $H_a: p < .50$; we can use a large sample test if both $np_0 \geq 10$ and $n(1 - p_0) \geq 10$; $102(.50) = 51$, so we can proceed. $\hat{p} = \frac{47}{102}$, so

$z = \dfrac{\frac{47}{102} - .50}{\sqrt{\frac{(.50)(.50)}{102}}} = \dfrac{-.039}{.050} = -.79$. We will reject H_0 if the P-value $< .01$. For this lower tailed test, the P-value $= \Phi(z) = \Phi(-.79) = .2148$, which is not $< .01$, so we do not reject H_0 at significance level .01.

57. The hypotheses to be tested are $H_0: \mu = 25$ v. $H_a: \mu > 25$. The computed summary statistics are $\bar{x} = 27.923$ and $s = 5.619$, so $\frac{s}{\sqrt{n}} = 1.559$ and $t = \frac{2.923}{1.559} = 1.88$. From table A.8, $P(t > 1.88) \approx .041$, which is less than .05, so H_0 is rejected at level .05.

59. μ = true average reading, $H_0: \mu = 70$ v. $H_a: \mu \neq 70$, and $t = \dfrac{\bar{x} - 70}{s / \sqrt{n}} = \dfrac{75.5 - 70}{7 / \sqrt{6}} =$ $\dfrac{5.5}{2.86} = 1.92$. From table A.8, df = 5, P-value $= 2[P(t > 1.92)] \approx 2(.058) = .116$. At significance level .05, there is not enough evidence to conclude that the spectrophotometer needs recalibrating.

Section 8.5

61.

    a. The formula for β is $1 - \Phi\left(-2.33 + \dfrac{\sqrt{n}}{9}\right)$, which gives .8888 for n = 100, .1587 for n = 900, and .0006 for n = 2500.

    b. Z = –5.3, which is "off the z table," so P-value < .0002; this value of z is quite statistically significant.

    c. No. Even when the departure from H_0 is insignificant from a practical point of view, a statistically significant result is highly likely to appear; the test is too likely to detect small departures from H_0.

Supplementary Exercises

63. Because n = 50 is large, we use a z test here, rejecting H_0: $\mu = 3.2$ in favor of H_a:

    $\mu \neq 3.2$ if either $z \geq 1.96$ or $z \leq -1.96$. The computed z value is $z = \dfrac{3.05 - 3.20}{.34 / \sqrt{50}} =$

    –3.12. Since $-3.12 \leq -1.96$, H_0 should be rejected in favor of H_a.

65.

    a. H_0: $\mu = .85$ v. H_a: $\mu \neq .85$

    b. With a P-value of .30, we would reject the null hypothesis at any reasonable significance level, which includes both .05 and .10.

67.

    a. The relevant hypotheses are H_0: $\mu = 548$ v. H_a: $\mu \neq 548$. At level .05, H_0 will be rejected if either $t \geq t_{.025,10} = 2.228$ or $t \leq -t_{.025,10} = -2.228$. The test statistic

    value is $t = \dfrac{587 - 548}{10 / \sqrt{11}} = \dfrac{39}{3.02} = 12.9$. This clearly falls into the upper tail of the

    two-tailed rejection region, so H_0 should be rejected at level .05, or any other reasonable level).

    b. The population sampled was normal or approximately normal.

69. $n = 47$, $\bar{x} = 215$ mg, $s = 235$ mg, range = 5 mg to 1,176 mg

    a. No, the distribution does not appear to be normal. It appears to be skewed to the right, since 0 is less than one standard deviation below the mean. It is not necessary to assume normality if the sample size is large enough due to the central limit theorem. This sample size is large enough so we can conduct a hypothesis test about the mean.

    b.

        1   Parameter of interest: μ = true daily caffeine consumption of adult women.

        2   H_0: $\mu = 200$

        3   H_a: $\mu > 200$

        4   $z = \dfrac{\bar{x} - 200}{s/\sqrt{n}}$

        5   RR: $z \geq 1.282$ or if P-value $\leq .10$

        6   $z = \dfrac{215 - 200}{235/\sqrt{47}} = .44$; P-value $= 1 - \Phi(.44) = .33$

        7   Fail to reject H_0. because $.33 > .10$. The data does not indicate that daily consumption of all adult women exceeds 200 mg.

71.

    a. From Table A.17, when $\mu = 9.5$, $d = .625$, and df = 9, $\beta \approx .60$. When $\mu = 9.0$, $d = 1.25$, and df = 9, $\beta \approx .20$.

    b. From Table A.17, when $\beta = .25$ and $d = .625$, $n \approx 28$

73.

    a. With H_0: $p = 1/75$ v. H_a: $p \neq 1/75$, we reject H_0 if either $z \geq 1.96$ or $z \leq -1.96$. With $\hat{p} = \dfrac{16}{800} = .02$, $z \dfrac{.02 - .01333}{\sqrt{\dfrac{.01333(.98667)}{800}}} = 1.645$, which is not in either rejection region. Thus, we fail to reject the null hypothesis. There is not evidence that the incidence rate among prisoners differs from that of the adult population. The possible error we could have made is a type II.

    b. P–value $= 2[1 - \Phi(1.645)] = 2[.05] = .10$. Yes, since $.10 < .20$, we could reject H_0.

75. Even though the underlying distribution may not be normal, a z test can be used because n is large. H_0: $\mu = 3200$ should be rejected in favor of H_a: $\mu < 3200$ if $z \le -z_{.001} = -3.08$. The computed test statistic is $z = \dfrac{3107 - 3200}{188/\sqrt{45}} = -3.32 \le -3.08$, so H_0 should be rejected at level .001.

77. We wish to test H_0: $\lambda = 4$ v. H_a: $\lambda > 4$ using the test statistic $z = \dfrac{\bar{x} - 4}{\sqrt{4/n}}$. For the given sample, $n = 36$ and $\bar{x} = \dfrac{160}{36} = 4.444$, so $z = \dfrac{4.444 - 4}{\sqrt{4/36}} = 1.33$. At level .02, we reject H_0 if $z \ge z_{.02} \approx 2.05$ (since $1 - \Phi(2.05) = .0202$). Because 1.33 is not ≥ 2.05, H_0 should not be rejected at this level.

79. H_0: $\mu = 15$ v. H_a: $\mu > 15$. Because the sample size is less than 40, and we can assume the distribution is approximately normal, the appropriate statistic is
$$t = \frac{\bar{x} - 15}{s/\sqrt{n}} = \frac{17.5 - 15}{2.2/\sqrt{32}} = \frac{2.5}{.390} = 6.4.$$ This t statistic is "off the chart" in the 20 df column of Table A.8, so P-value $\approx 0 < .05$, and H_0 is rejected in favor of the conclusion that the true average time exceeds 15 minutes.

81. The 20 df row of Table A.7 shows that $\chi^2_{.99,20} = 8.26 < 8.58$ (H_0 not rejected at level .01) and $8.58 < 9.591 = \chi^2_{.975,20}$ (H_0 rejected at level .025). Thus $.01 < P\text{-value} < .025$ and H_0 cannot be rejected at level .01 (the P-value is the smallest α at which rejection can take place, and this exceeds .01).

Chapter 8: Tests of Hypotheses Based on a Single Sample

83.

a. When H_0 is true, $2\lambda_o \Sigma X_i = 2\sum \dfrac{X_i}{\mu_o}$ has a chi-squared distribution with df $= 2n$.

If the alternative is H_a: $\mu > \mu_o$, large test statistic values (large Σx_i, since $\bar{x}$ is large) suggest that H_0 be rejected in favor of H_a, so rejecting when

$2\sum \dfrac{X_i}{\mu_o} \geq \chi^2_{\alpha,2n}$ gives a test with significance level α. If the alternative is H_a:

$\mu < \mu_o$, rejecting when $2\sum \dfrac{X_i}{\mu_o} \leq \chi^2_{1-\alpha,2n}$ gives a level α test. The rejection

region for H_a: $\mu \neq \mu_o$ is either $2\sum \dfrac{X_i}{\mu_o} \geq \chi^2_{\alpha/2,2n}$ or $\leq \chi^2_{1-\alpha/2,2n}$.

b. H_0: $\mu = 75$ v. H_a: $\mu < 75$. The test statistic value is $\dfrac{2(737)}{75} = 19.65$. At level

.01, H_0 is rejected if $2\sum \dfrac{X_i}{\mu_o} \leq \chi^2_{.99,20} = 8.260$. Clearly 19.65 is not in the

rejection region, so H_0 should not be rejected. The sample data does not suggest that true average lifetime is less than the previously claimed value.

85.

a. $\alpha = P(X \leq 5 \text{ when } p = .9) = B(5; 10, .9) = .002$, so the region $(0, 1, ..., 5)$ does specify a level .01 test.

b. The first value to be placed in the upper-tailed part of a two tailed region would be 10, but $P(X = 10 \text{ when } p = .9) = .349$, so whenever 10 is in the rejection region, $\alpha \geq .349$.

c. $\beta(p') = P(X \text{ is } \underline{not} \text{ in } R \text{ when } p = p') = P(X > 5 \text{ when } p = p') = 1 - B(5; 10, p')$. The test has no ability to detect a false null hypothesis when $p > .90$ (see the graph for $.90 < p' < 1$). This is a by-product of the unavoidable one-sided rejection region (see **a** and **b**). The test also has an undesirably high $\beta(p')$ for medium-to-large p', a result of the small sample size.

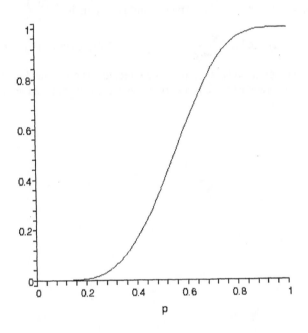

CHAPTER 9

Section 9.1

1.

    a. $E(\bar{X} - \bar{Y}) = E(\bar{X}) - E(\bar{Y}) = 4.1 - 4.5 = -.4$, irrespective of sample sizes.

    b. $V(\bar{X} - \bar{Y}) = V(\bar{X}) + V(\bar{Y}) = \dfrac{\sigma_1^2}{m} + \dfrac{\sigma_2^2}{n} = \dfrac{(1.8)^2}{100} + \dfrac{(2.0)^2}{100} = .0724$, and the standard deviation of $\bar{X} - \bar{Y} = \sqrt{.0724} = .2691$.

    c.   A normal curve with mean and s.d. as given in **a** and **b** (because m = n = 100, the CLT implies that both $\bar{X}$ and $\bar{Y}$ have approximately normal distributions, so $\bar{X} - \bar{Y}$ does also). The shape is not necessarily that of a normal curve when m = n = 10, because the CLT cannot be invoked. So if the two lifetime population distributions are not normal, the distribution of $\bar{X} - \bar{Y}$ will typically be quite complicated.

3.   The test statistic value is $z = \dfrac{(\bar{x} - \bar{y}) - 5000}{\sqrt{\dfrac{s_1^2}{m} + \dfrac{s_2^2}{n}}}$, and H$_0$ will be rejected at level .01 if

$z \geq 2.33$. We compute $z = \dfrac{(42{,}500 - 36{,}800) - 5000}{\sqrt{\dfrac{2200^2}{45} + \dfrac{1500^2}{45}}} = \dfrac{700}{396.93} = 1.76$, which is less

than 2.33, so we don't reject H$_0$ and conclude that the true average life for radials does not exceed that for economy brand by significantly more than 500.

Chapter 9: Inferences Based on Two Samples

5.

a. H_a says that the average calorie output for sufferers is more than 1 cal/cm²/min below that for non-sufferers. $\sqrt{\dfrac{\sigma_1^2}{m} + \dfrac{\sigma_2^2}{n}} = \sqrt{\dfrac{(.2)^2}{10} + \dfrac{(.4)^2}{10}} = .1414$, so

$z = \dfrac{(.64 - 2.05) - (-1)}{.1414} = -2.90$. At level .01, H_0 is rejected if $z \le -2.33$; since $-2.90 < -2.33$, reject H_0.

b. $P = \Phi(-2.90) = .0019$

c. $\beta = 1 - \Phi\left(-2.33 - \dfrac{-1.2 + 1}{.1414}\right) = 1 - \Phi(-.92) = .8212$

d. $m = n = \dfrac{.2(2.33 + 1.28)^2}{(-.2)^2} = 65.15$, so use 66.

7.

1.   Parameter of interest: $\mu_1 - \mu_2$ = the true difference of means for males and females on the Boredom Proneness Rating. Let μ_1 = men's average and μ_2 = women's average.

2. $H_0 : \mu_1 - \mu_2 = 0$

3. $H_a : \mu_1 - \mu_2 > 0$

4. $z = \dfrac{(\bar{x} - \bar{y}) - \Delta_o}{\sqrt{\dfrac{s_1^2}{m} + \dfrac{s_2^2}{n}}} = \dfrac{(\bar{x} - \bar{y}) - 0}{\sqrt{\dfrac{s_1^2}{m} + \dfrac{s_2^2}{n}}}$

5.   RR: $z \ge 1.645$

6. $z = \dfrac{(10.40 - 9.26) - 0}{\sqrt{\dfrac{4.83^2}{97} + \dfrac{4.68^2}{148}}} = 1.83$

7.   Reject H_0. The data indicates the average Boredom Proneness Rating is higher for males than for females.

Chapter 9: Inferences Based on Two Samples

9.

    a. point estimate $\bar{x} - \bar{y} = 19.9 - 13.7 = 6.2$. It appears that there could be a difference.

    b.

$$H_0: \mu_1 - \mu_2 = 0, H_a: \mu_1 - \mu_2 \neq 0, \ z = \frac{(19.9 - 13.7)}{\sqrt{\dfrac{39.1^2}{60} + \dfrac{15.8^2}{60}}} = \frac{6.2}{5.44} = 1.14, \text{ and the}$$

p-value $= 2[P(Z > 1.14)] = 2(.1271) = .2542$. The P-value is larger than any reasonable α, so we do not reject H_0. There is no significant difference.

    c. No. With a normal distribution, we would expect most of the data to be within 2 standard deviations of the mean, and the distribution should be symmetric. 2 sd's above the mean is 98.1, but the distribution stops at zero on the left. The distribution is positively skewed.

    d. We will calculate a 95% confidence interval for μ, the true average length of stays for patients given the treatment: $19.9 \pm 1.96 \dfrac{39.1}{\sqrt{60}} = 19.9 \pm 9.9 = (10.0, 21.8)$

11. $(\bar{X} - \bar{Y}) \pm z_{\alpha/2} \sqrt{\dfrac{s_1^2}{m} + \dfrac{s_2^2}{n}}$. Standard error $= \dfrac{s}{\sqrt{n}}$. Substitution yields

$(\bar{x} - \bar{y}) \pm z_{\alpha/2} \sqrt{(SE_1)^2 + (SE_2)^2}$. Using $\alpha = .05$, $z_{\alpha/2} = 1.96$, so

$(5.5 - 3.8) \pm 1.96 \sqrt{(0.3)^2 + (0.2)^2} = (0.99, 2.41)$. We are 95% confident that the true average blood lead level for male workers is between 0.99 and 2.41 higher than the corresponding average for female workers.

13. $\sigma_1 = \sigma_2 = .05$, $d = .04$, $\alpha = .01, \beta = .05$, and the test is one-tailed, so

$$n = \frac{(.0025 + .0025)(2.33 + 1.645)^2}{.0016} = 49.38, \text{ so use } n = 50.$$

15.

a. As either m or n increases, σ decreases, so $\dfrac{\mu_1 - \mu_2 - \Delta_o}{\sigma}$ increases (the

numerator is positive), so $\left(z_\alpha - \dfrac{\mu_1 - \mu_2 - \Delta_o}{\sigma} \right)$ decreases, so

$\beta = \Phi\left(z_\alpha - \dfrac{\mu_1 - \mu_2 - \Delta_o}{\sigma} \right)$ decreases.

b. As β decreases, z_β increases; since z_β is the numerator of n, n increases also.

Section 9.2

17.

a. $v = \dfrac{\left(\frac{5^2}{10} + \frac{6^2}{10} \right)^2}{\dfrac{\left(\frac{5^2}{10} \right)^2}{9} + \dfrac{\left(\frac{6^2}{10} \right)^2}{9}} = \dfrac{37.21}{.694 + 1.44} = 17.43 \approx 17$

b. $v = \dfrac{\left(\frac{5^2}{10} + \frac{6^2}{15} \right)^2}{\dfrac{\left(\frac{5^2}{10} \right)^2}{9} + \dfrac{\left(\frac{6^2}{15} \right)^2}{14}} = \dfrac{24.01}{.694 + .411} = 21.7 \approx 21$

c. $v = \dfrac{\left(\frac{2^2}{10} + \frac{6^2}{15} \right)^2}{\dfrac{\left(\frac{2^2}{10} \right)^2}{9} + \dfrac{\left(\frac{6^2}{15} \right)^2}{14}} = \dfrac{7.84}{.018 + .411} = 18.27 \approx 18$

d. $v = \dfrac{\left(\frac{5^2}{12} + \frac{6^2}{24} \right)^2}{\dfrac{\left(\frac{5^2}{12} \right)^2}{11} + \dfrac{\left(\frac{6^2}{24} \right)^2}{23}} = \dfrac{12.84}{.395 + .098} = 26.05 \approx 26$

19. For the given hypotheses, the test statistic $t = \dfrac{115.7 - 129.3 + 10}{\sqrt{\frac{5.03^2}{6} + \frac{5.38^2}{6}}} = \dfrac{-3.6}{3.007} = -1.20$,

and the d.f. is $\nu = \dfrac{(4.2168 + 4.8241)^2}{\dfrac{(4.2168)^2}{5} + \dfrac{(4.8241)^2}{5}} = 9.96$, so use d.f. = 9. We will reject H_0 if

$t \le -t_{.01,9} = -2.764$; since $-1.20 > -2.764$, we don't reject H_0.

21. Let $\mu_1 =$ the true average gap detection threshold for normal subjects, and $\mu_2 =$ the corresponding value for CTS subjects. The relevant hypotheses are $H_0: \mu_1 - \mu_2 = 0$

vs. $H_a: \mu_1 - \mu_2 < 0$, and the test statistic $t = \dfrac{1.71 - 2.53}{\sqrt{.0351125 + .07569}} = \dfrac{-.82}{.3329} = -2.46$.

Using d.f. $\nu = \dfrac{(.0351125 + .07569)^2}{\dfrac{(.0351125)^2}{7} + \dfrac{(.07569)^2}{9}} = 15.1$, or 15, the rejection region is

$t \le -t_{.01,15} = -2.602$. Since -2.46 is not ≤ -2.602, we fail to reject H_0. We have insufficient evidence to claim that the true average gap detection threshold for CTS subjects exceeds that for normal subjects.

23.

a. We see that both plots illustrate sufficient linearity. Therefore, it is plausible that both samples have been selected from normal population distributions.

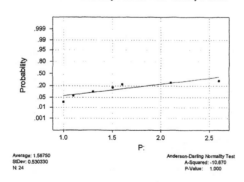

Normal Probability Plot for Poor Quality Fabric

Normal Probability Plot for High Quality Fabric

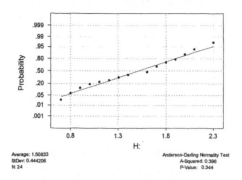

Average: 1.50833
StDev: 0.444206
N: 24

Anderson-Darling Normality Test
A-Squared: 0.396
P-Value: 0.344

b. The comparative boxplot does not suggest a difference between average extensibility for the two types of fabrics.

Comparative Box Plot for High Quality and Poor Quality Fabric

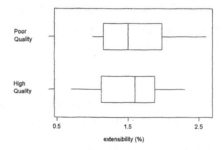

c. We test $H_0 : \mu_1 - \mu_2 = 0$ vs. $H_a : \mu_1 - \mu_2 \neq 0$. With degrees of freedom $\nu = \dfrac{(.0433265)^2}{.00017906} = 10.5$, which we round down to 10, and using significance level .05 (not specified in the problem), we reject H_0 if $|t| \geq t_{.025,10} = 2.228$. The test statistic is $t = \dfrac{-.08}{\sqrt{(.0433265)}} = -.38$, which is not ≥ 2.228 in absolute value, so we cannot reject H_0. There is insufficient evidence to claim that the true average extensibility differs for the two types of fabrics.

25.   We calculate the degrees of freedom $\nu = \dfrac{\left(\frac{5.5^2}{28} + \frac{7.8^2}{31}\right)^2}{\frac{\left(\frac{5.5^2}{28}\right)^2}{27} + \frac{\left(\frac{7.8^2}{31}\right)^2}{30}} = 53.95$, or about 54

(normally we would round down to 53, but this number is very close to 54 – of course for this large number of df, using either 53 or 54 won't make much difference in the critical t value) so the desired confidence interval is $(91.5 - 88.3) \pm 1.68\sqrt{\frac{5.5^2}{28} + \frac{7.8^2}{31}}$

$= 3.2 \pm 2.931 = (.269, 6.131)$. Because 0 does not lie inside this interval, we can be reasonably certain that the true difference $\mu_1 - \mu_2$ is not 0 and, therefore, that the two population means are not equal. For a 95% interval, the t value increases to about 2.01 or so, which results in the interval 3.2 ± 3.506. Since this interval does contain 0, we can no longer conclude that the means are different if we use a 95% confidence interval.

27.   The approximate degrees of freedom for this estimate are

$\nu = \dfrac{\left(\frac{11.3^2}{6} + \frac{8.3^2}{8}\right)^2}{\frac{\left(\frac{11.3^2}{6}\right)^2}{5} + \frac{\left(\frac{8.3^2}{8}\right)^2}{7}} = \dfrac{893.59}{101.175} = 8.83$, which we round down to 8, so $t_{.025,8} = 2.306$

and the desired interval is $(40.3 - 21.4) \pm 2.306\sqrt{\frac{11.3^2}{6} + \frac{8.3^2}{8}} = 18.9 \pm 2.306(5.4674)$

$= 18.9 \pm 12.607 = (6.3, 31.5)$. Because 0 is not contained in this interval, there is strong evidence that $\mu_1 - \mu_2$ is not 0; i.e., we can conclude that the population means are not equal. Calculating a confidence interval for $\mu_2 - \mu_1$ would change only the order of subtraction of the sample means, but the standard error calculation would give the same result as before. Therefore, the 95% interval estimate of $\mu_2 - \mu_1$ would be (–31.5, –6.3), just the negatives of the endpoints of the original interval. Since 0 is not in this interval, we reach exactly the same conclusion as before; the population means are not equal.

29. Let μ_1 = the true average compression strength for strawberry drink and let μ_2 = the true average compression strength for cola. A lower tailed test is appropriate. We test $H_0 : \mu_1 - \mu_2 = 0$ vs. $H_a : \mu_1 - \mu_2 < 0$. The test statistic is

$$t = \frac{-14}{\sqrt{29.4 + 15}} = -2.10 \, . \quad v = \frac{(44.4)^2}{\frac{(29.4)^2}{14} + \frac{(15)^2}{14}} = \frac{1971.36}{77.8114} = 25.3 \text{, so use df} = 25.$$

The P-value $\approx P(t < -2.10) = .023$. This P-value indicates strong support for the alternative hypothesis. The data does suggest that the extra carbonation of cola results in a higher average compression strength.

31.

a. The most notable feature of these boxplots is the larger amount of variation present in the mid-range data compared to the high-range data. Otherwise, both look reasonably symmetric with no outliers present.

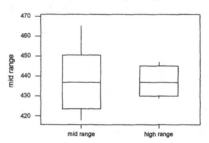

Comparative Box Plot for High Range and Mid Range

b. Using df = 23, a 95% confidence interval for $\mu_{mid-range} - \mu_{high-range}$ is

$$\left(438.3 - 437.45\right) \pm 2.069 \sqrt{\frac{15.1^2}{17} + \frac{6.83^2}{11}} = .85 \pm 8.69 = \left(-7.84, 9.54\right). \text{ Since}$$

plausible values for $\mu_{mid-range} - \mu_{high-range}$ are both positive and negative (i.e., the interval spans zero) we would conclude that there is not sufficient evidence to suggest that the average value for mid-range and the average value for high-range differ.

33. Let μ_1 and μ_2 represent the true mean body mass decrease for the vegan diet and the control diet, respectively. We wish to test $H_0: \mu_1 - \mu_2 \le 1$ v. $H_a: \mu_1 - \mu_2 > 1$. The

relevant test statistic is $t = \dfrac{(5.8 - 3.8) - 1}{\sqrt{\dfrac{3.2^2}{32} + \dfrac{2.8^2}{32}}} = 1.33$, with estimated df $= 60$ using the

formula. Rounding to $t = 1.3$, Table A.8 gives a one-sided P-value of .098 (a computer will give the more accurate P-value of .094). Since our P-value $> \alpha = .05$, we fail to reject H_0 at the 5% level. We do not have statistically significant evidence that the true average weight loss for the vegan diet exceeds that for the control diet by more than 1 kg.

35. There are two changes that must be made to the procedure we currently use. First, the equation used to compute the value of the t test statistic is: $t = \dfrac{(\bar{x} - \bar{y}) - (\Delta)}{s_p \sqrt{\dfrac{1}{m} + \dfrac{1}{n}}}$ where s_p

is defined as in Exercise 34 above. Second, the degrees of freedom $= m + n - 2$. Assuming equal variances in the situation from Exercise 33, we calculate s_p as

follows: $s_p = \sqrt{\left(\dfrac{7}{16}\right)(2.6)^2 + \left(\dfrac{9}{16}\right)(2.5)^2} = 2.544$. The value of the test statistic is,

then, $t = \dfrac{(32.8 - 40.5) - (-5)}{2.544\sqrt{\dfrac{1}{8} + \dfrac{1}{10}}} = -2.24 \approx -2.2$. The degrees of freedom $= 16$, and the

p-value is $P(t < -2.2) = .021$. Since $.021 > .01$, we fail to reject H_0.

Section 9.3

37.

    a. This exercise calls for paired analysis. First, compute the difference between indoor and outdoor concentrations of hexavalent chromium for each of the 33 houses. These 33 differences are summarized as follows: $n = 33$, $\bar{d} = -.4239$, $s_d = .3868$, where $d =$ (indoor value – outdoor value). Then $t_{.025,32} = 2.037$, and a 95% confidence interval for the population mean difference between indoor and outdoor concentration is

$$-.4239 \pm (2.037)\left(\frac{.3868}{\sqrt{33}}\right) = -.4239 \pm .13715 = (-.5611, -.2868).$$ We can be

highly confident, at the 95% confidence level, that the true average concentration of hexavalent chromium outdoors exceeds the true average concentration indoors by between .2868 and .5611 nanograms/m^3.

    b. A 95% prediction interval for the difference in concentration for the 34th house is

$$\bar{d} \pm t_{.025,32}\left(s_d \sqrt{1 + \tfrac{1}{n}}\right) = -.4239 \pm (2.037)\left(.3868\sqrt{1 + \tfrac{1}{33}}\right) = (-1.224, .3758).$$ This

prediction interval means that the indoor concentration may exceed the outdoor concentration by as much as .3758 nanograms/m^3 and that the outdoor concentration may exceed the indoor concentration by a much as 1.224 nanograms/m^3, for the 34th house. Clearly, this is a wide prediction interval, largely because of the amount of variation in the differences.

39.

    a. A normal probability plot shows that the data could easily follow a normal distribution.

    b. We test $H_0 : \mu_d = 0$ vs. $H_a : \mu_d \neq 0$, with test statistic

$$t = \frac{\bar{d} - 0}{s_D / \sqrt{n}} = \frac{167.2 - 0}{228 / \sqrt{14}} = 2.74 \approx 2.7.$$ The two-tailed p-value is $2[P(t > 2.7)] =$

$2[.009] = .018$. Since $.018 < .05$, we reject H_0. There is strong evidence to support the claim that the true average difference between intake values measured by the two methods is not 0. There is a difference between them.

41. We test $H_0 : \mu_d = 5$ vs. $H_a : \mu_d > 5$. With $\overline{d} = 7.600$, and $s_d = 4.178$,

$t = \dfrac{7.600 - 5}{4.178/\sqrt{9}} = \dfrac{2.6}{1.39} = 1.87 \approx 1.9$. With degrees of freedom $n - 1 = 8$, the

corresponding p-value is $P(t > 1.9) = .047$. We would reject H_0 at any alpha level greater than .047. So, at the typical significance level of .05, we would reject H_0, and conclude that the data indicates that the higher level of illumination yields a decrease of more than 5 seconds in true average task completion time.

43.

    a. Although there is a "jump" in the middle of the Normal Probability plot, the data follow a reasonably straight path, so there is no strong reason for doubting the normality of the population of differences.

    b. A 95% lower confidence bound for the population mean difference is:

$$\overline{d} - t_{.05,14}\left(\frac{s_d}{\sqrt{n}}\right) = -38.60 - (1.761)\left(\frac{23.18}{\sqrt{15}}\right) = -38.60 - 10.54 = -49.14 \text{ We are}$$

95% confident that the true mean difference between age at onset of Cushing's disease symptoms and age at diagnosis is greater than -49.14.

    c. A 95% upper confidence bound for the corresponding population mean difference is $38.60 + 10.54 = 49.14$.

45. From the data, $n = 12$, $\overline{d} = -0.73$, $s_d = 2.81$.

    a. Let μ_d = the true mean difference in strength between curing under moist conditions and laboratory drying conditions. A 95% CI for μ_d is $\overline{d} \pm t_{.025,11}s_d/\sqrt{n}$

$= -0.73 \pm 2.201(2.81)/\sqrt{10} = (-2.52 \text{ MPa}, 1.05 \text{ MPa})$. In particular, this interval estimate includes the value zero, suggesting that true mean strength is not significantly different under these two conditions.

    b. Since $n = 12$, we must check that the <u>differences</u> are plausibly from a normal population. The normal probability plot (see next page) strongly substantiates that condition.

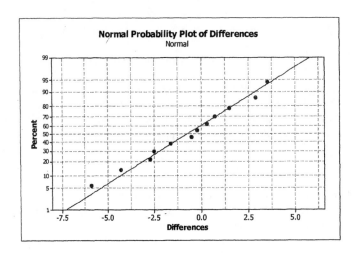

Section 9.4

47. H_0 will be rejected if $z \le -z_{.01} = -2.33$. With $\hat{p}_1 = .150$, and $\hat{p}_2 = .300$,

$\hat{p} = \dfrac{30 + 80}{200 + 600} = \dfrac{210}{800} = .263$, and $\hat{q} = .737$. The calculated test statistic is

$z = \dfrac{.150 - .300}{\sqrt{(.263)(.737)\left(\frac{1}{200} + \frac{1}{600}\right)}} = \dfrac{-.150}{.0359} = -4.18$. Because $-4.18 \le -2.33$, H_0 is

rejected; the proportion of those who repeat after inducement appears lower than those who repeat after no inducement.

49.

1   Parameter of interest: $p_1 - p_2 =$ true difference in proportions of those responding to two different survey covers. Let $p_1 =$ Plain, $p_2 =$ Picture.

2     $H_0 : p_1 - p_2 = 0$

3     $H_a : p_1 - p_2 < 0$

4     $z = \dfrac{\hat{p}_1 - \hat{p}_2}{\sqrt{\hat{p}\hat{q}\left(\frac{1}{m} + \frac{1}{n}\right)}}$

5   Reject H_0 if p-value $< .10$

6     $z = \dfrac{\frac{104}{207} - \frac{109}{213}}{\sqrt{\left(\frac{213}{420}\right)\left(\frac{207}{420}\right)\left(\frac{1}{207} + \frac{1}{213}\right)}} = -.1910$; p-value $= .4247$

7   Fail to Reject H_0. The data does not indicate that plain cover surveys have a lower response rate.

Chapter 9: Inferences Based on Two Samples

51.

a. Let p_1 and p_2 denote the true incidence rates of GI problems for the olestra and control groups, respectively. We wish to test $H_0: p_1 - \mu_2 = 0$ v. $H_a: p_1 - p_2 \neq 0$.

The pooled proportion is $\hat{p} = \dfrac{529(.176) + 563(.158)}{529 + 563} = .1667$, from which the

relevant test statistic is $z = \dfrac{.176 - .158}{\sqrt{(.1667)(.8333)[529^{-1} + 563^{-1}]}} = 0.78$. The two-

sided P-value is $2P(Z \geq 0.78) = .433 > \alpha = .05$, hence we fail to reject the null hypothesis. The data do not suggest a statistically significant difference between the incidence rates of GI problems between the two groups.

b. $n = \dfrac{\left(1.96\sqrt{(.35)(1.65)/2} + 1.28\sqrt{(.15)(.85) + (.2)(.8)}\right)^2}{(.05)^2} = 1210.39$, so a common

sample size of $m = n = 1211$ would be required.

53.

a. A 95% large sample confidence interval formula for $\ln(\theta)$ is

$\ln(\hat{\theta}) \pm z_{\alpha/2} \sqrt{\dfrac{m-x}{mx} + \dfrac{n-y}{ny}}$. Taking the antilogs of the upper and lower bounds

gives the confidence interval for θ itself.

b. $\hat{\theta} = \dfrac{\frac{189}{11,034}}{\frac{104}{11,037}} = 1.818$, $\ln(\hat{\theta}) = .598$, and the standard deviation is

$\sqrt{\dfrac{10,845}{(11,034)(189)} + \dfrac{10,933}{(11,037)(104)}} = .1213$, so the CI for $\ln(\theta)$ is

$.598 \pm 1.96(.1213) = (.360, .836)$. Then taking the antilogs of the two bounds gives the CI for θ to be (1.43, 2.31). We are 95% confident that people who do not take the aspirin treatment are between 1.43 and 2.31 times more likely to suffer a heart attack than those who do. This suggests aspirin therapy may be effective in reducing the risk of a heart attack.

55. $\hat{p}_1 = \dfrac{15 + 7}{40} = .550$, $\hat{p}_2 = \dfrac{29}{42} = .690$, and the 95% C.I. is

$(.550 - .690) \pm 1.96(.106) = -.14 \pm .21 = (-.35, .07)$.

Section 9.5

57.

 a. From Table A.9, column 5, row 8, $F_{.01,5,8} = 3.69$.

 b. From column 8, row 5, $F_{.01,8,5} = 4.82$.

 c. $F_{.95,5,8} = \dfrac{1}{F_{.05,8,5}} = .207$.

 d. $F_{.95,8,5} = \dfrac{1}{F_{.05,5,8}} = .271$

 e. $F_{.01,10,12} = 4.30$

 f. $F_{.99,10,12} = \dfrac{1}{F_{.01,12,10}} = \dfrac{1}{4.71} = .212$.

 g. $F_{.05,6,4} = 6.16$, so $P(F \le 6.16) = .95$.

 h. Since $F_{.99,10,5} = \dfrac{1}{5.64} = .177$, $P(.177 \le F \le 4.74) = P(F \le 4.74) - P(F \le .177)$

  $= .95 - .01 = .94$.

59. We test $H_0 : \sigma_1^2 = \sigma_2^2$ vs. $H_a : \sigma_1^2 \ne \sigma_2^2$. The calculated test statistic is

$f = \dfrac{(2.75)^2}{(4.44)^2} = .384$. With numerator d.f. $= m - 1 = 10 - 1 = 9$, and denominator d.f.

$= n - 1 = 5 - 1 = 4$, we reject H_0 if $f \ge F_{.05,9,4} = 6.00$ or

$f \le F_{.95,9,4} = \dfrac{1}{F_{.05,4,9}} = \dfrac{1}{3.63} = .275$. Since .384 is in neither rejection region, we

do not reject H_0 and conclude that there is no significant difference between the two standard deviations.

61. Let σ_1^2 = variance in weight gain for low-dose treatment, and σ_2^2 = variance in weight gain for control condition. We wish to test $H_0 : \sigma_1^2 = \sigma_2^2$ vs. $H_a : \sigma_1^2 > \sigma_2^2$. The test statistic is $f = \dfrac{s_1^2}{s_2^2}$, and we reject H_0 at level .05 if $f > F_{.05,19,22} \approx 2.08$.

$f = \dfrac{(54)^2}{(32)^2} = 2.85 \geq 2.08$, so reject H_0 at level .05. The data does suggest that there is more variability in the low-dose weight gains.

63. $P\left(F_{1-\alpha/2,m-1,n-1} \leq \dfrac{S_1^2 / \sigma_1^2}{S_2^2 / \sigma_2^2} \leq F_{\alpha/2,m-1,n-1} \right) = 1-\alpha$. The set of inequalities inside the parentheses is clearly equivalent to $\dfrac{S_2^2 F_{1-\alpha/2,m-1,n-1}}{S_1^2} \leq \dfrac{\sigma_2^2}{\sigma_1^2} \leq \dfrac{S_2^2 F_{\alpha/2,m-1,n-1}}{S_1^2}$.

Substituting the sample values s_1^2 and s_2^2 yields the confidence interval for $\dfrac{\sigma_2^2}{\sigma_1^2}$, and taking the square root of each endpoint yields the confidence interval for $\dfrac{\sigma_2}{\sigma_1}$. $m = n$

= 4, so we need $F_{.05,3,3} = 9.28$ and $F_{.95,3,3} = \dfrac{1}{9.28} = .108$. Then with $s_1 = .160$ and $s_2 = .074$, the C. I. for $\dfrac{\sigma_2^2}{\sigma_1^2}$ is (.023, 1.99), and for $\dfrac{\sigma_2}{\sigma_1}$ is (.15, 1.41).

Supplementary Exercises

65. We test $H_0 : \mu_1 - \mu_2 = 0$ vs. $H_a : \mu_1 - \mu_2 \neq 0$. The test statistic is

$$t = \frac{(\bar{x} - \bar{y}) - (\Delta)}{\sqrt{\frac{s_1^2}{m} + \frac{s_2^2}{n}}} = \frac{807 - 757}{\sqrt{\frac{27^2}{10} + \frac{41^2}{10}}} = \frac{50}{\sqrt{241}} = \frac{50}{15.524} = 3.22 \, . \text{ The approximate d.f. is}$$

$$\nu = \frac{(241)^2}{\frac{(72.9)^2}{9} + \frac{(168.1)^2}{9}} = 15.6 \, , \text{ which we round down to 15. The p-value for a two-}$$

tailed test is approximately $2P(t > 3.22) = 2(.003) = .006$. This small of a p-value gives strong support for the alternative hypothesis. The data indicates a significant difference. Due to the small sample sizes (10 each), we are assuming here that compression strengths for both fixed and floating test platens are normally distributed. And, as always, we are assuming the data were randomly sampled from their respective populations.

67. Let p_1 = true proportion of returned questionnaires that included no incentive; p_2 = true proportion of returned questionnaires that included an incentive. The hypotheses are $H_0 : p_1 - p_2 = 0$ vs. $H_0 : p_1 - p_2 < 0$. The test statistic is $z = \dfrac{\hat{p}_1 - \hat{p}_2}{\sqrt{\hat{p}\hat{q}\left(\frac{1}{m} + \frac{1}{n}\right)}}$.

$\hat{p}_1 = \dfrac{75}{110} = .682$, and $\hat{p}_2 = \dfrac{66}{98} = .673$. At this point we notice that since $\hat{p}_1 > \hat{p}_2$, the numerator of the z statistic will be > 0, and since we have a lower tailed test, the P-value will be $> .5$. We fail to reject H_0. This data does not suggest that including an incentive increases the likelihood of a response.

Chapter 9: Inferences Based on Two Samples

69.   The center of any confidence interval for $\mu_1 - \mu_2$ is always $\bar{x}_1 - \bar{x}_2$, so

$\bar{x}_1 - \bar{x}_2 = \dfrac{-473.3 + 1691.9}{2} = 609.3$. Furthermore, half of the width of this interval is

$\dfrac{1691.9 - (-473.3)}{2} = 1082.6$. Equating this value to the expression on the right of the

95% confidence interval formula, we find $\sqrt{\dfrac{s_1^2}{n_1} + \dfrac{s_2^2}{n_2}} = \dfrac{1082.6}{1.96} = 552.35$. For a 90%

interval, the associated z value is 1.645, so the 90% confidence interval is then
$609.3 \pm (1.645)(552.35) = 609.3 \pm 908.6 = (-299.3, 1517.9)$.

71. $m = n = 40$, $\bar{x} = 3975.0$, $s_1 = 245.1$, $\bar{y} = 2795.0$, $s_2 = 293.7$. The large sample 99%

confidence interval for $\mu_1 - \mu_2$ is $(3975.0 - 2795.0) \pm 2.58 \sqrt{\dfrac{245.1^2}{40} + \dfrac{293.7^2}{40}} =$

$(1180.0) \pm 1560 \approx (1020, 1340)$. The value 0 is not contained in this interval so we can
state that, with very high confidence, the value of $\mu_1 - \mu_2$ is not 0, which is
equivalent to concluding that the population means are not equal.

73.   Since we can assume that the distributions from which the samples were taken are
normal, we use the two-sample t test. Let μ_1 denote the true mean headability rating
for aluminum killed steel specimens and μ_2 denote the true mean headability rating
for silicon killed steel. Then the hypotheses are $H_0 : \mu_1 - \mu_2 = 0$ vs.

$H_a : \mu_1 - \mu_2 \neq 0$. The test statistic is $t = \dfrac{-.66}{\sqrt{.03888 + .047203}} = \dfrac{-.66}{\sqrt{.086083}} = -2.25$.

The approximate degrees of freedom $\nu = \dfrac{(.086083)^2}{\dfrac{(.03888)^2}{29} + \dfrac{(.047203)^2}{29}} = 57.5$, so we use

57. The two-tailed p-value $\approx 2(.014) = .028$, which is less than the specified
significance level, so we would reject H_0. The data supports the article's authors'
claim.

75.

a. The relevant hypotheses are $H_0 : \mu_1 - \mu_2 = 0$ vs. $H_a : \mu_1 - \mu_2 \neq 0$. Assuming both populations have normal distributions, the two-sample t test is appropriate. $m = 11$, $\bar{x} = 98.1$, $s_1 = 14.2$, $n = 15$, $\bar{y} = 129.2$, $s_2 = 39.1$. The test statistic is

$$t = \frac{-31.1}{\sqrt{18.3309 + 101.9207}} = \frac{-31.1}{\sqrt{120.252}} \doteq -2.84.$$ The approximate degrees of

freedom $\nu = \dfrac{(120.252)^2}{\dfrac{(18.3309)^2}{10} + \dfrac{(101.9207)^2}{14}} = 18.64$, so we use 18. From Table A.8,

the two-tailed p-value $\approx 2(.006) = .012$. No, obviously, the results are different.

b. For the hypotheses $H_0 : \mu_1 - \mu_2 = -25$ vs. $H_a : \mu_1 - \mu_2 < -25$, the test statistic

changes to $t = \dfrac{-31.1 - (-25)}{\sqrt{120.252}} = -.556$. With degrees of freedom 18, the p-value

$\approx P(t < -.6) = .278$. Since the p-value is greater than any sensible choice of α, we fail to reject H_0. There is insufficient evidence that the true average strength for males exceeds that for females by more than 25N.

77. This is paired data, so the paired t test is employed. The relevant hypotheses are $H_0 : \mu_d = 0$ vs. $H_a : \mu_d < 0$, where μ_d denotes the difference between the population average control strength minus the population average heated strength. The observed differences (control – heated) are: –.06, .01, –.02, 0, and –.05. The sample mean and standard deviation of the differences are $\bar{d} = -.024$ and

$s_d = .0305$. The test statistic is $t = \dfrac{-.024}{.0305 / \sqrt{5}} = -1.76 \approx -1.8$. From Table A.8,

with d.f. $= 5 - 1 = 4$, the lower tailed p-value associated with $t = -1.8$ is $P(t < -1.8) = P(t > 1.8) = .073$. At significance level .05, H_0 should not be rejected. Therefore, this data does not show that the heated average strength exceeds the average strength for the control population.

79. The normal probability plot below indicates the data for good visibility does not follow a normal distribution, thus a t-test is not appropriate for this small a sample size. (The plot for poor visibility isn't as bad.)

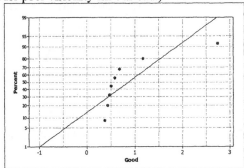

81. We wish to test H_0: $\mu_1 = \mu_2$ versus H_a: $\mu_1 \neq \mu_2$

Unpooled:

With H_0: $\mu_1 - \mu_2 = 0$ vs. H_a: $\mu_1 - \mu_2 \neq 0$; we will reject H_0 if $p - value < \alpha$.

$$\nu = \frac{\left(\frac{.79^2}{14} + \frac{1.52^2}{12}\right)^2}{\frac{\left(\frac{.79^2}{14}\right)^2}{13} + \frac{\left(\frac{1.52^2}{12}\right)^2}{11}} = 15.95 \downarrow 15 \text{, and the test statistic}$$

$$t = \frac{8.48 - 9.36}{\sqrt{\frac{.79^2}{14} + \frac{1.52^2}{12}}} = \frac{-.88}{.4869} = -1.81 \text{ leads to a p-value of about } 2[P(t_{15} > 1.8)] = 2(.046)$$

$= .092$.

Pooled:

The degrees of freedom are $\nu = m + n - 2 = 14 + 12 - 2 = 24$ and the pooled variance is $\left(\frac{13}{24}\right)(.79)^2 + \left(\frac{11}{24}\right)(1.52)^2 = 1.3970$, so $s_p = 1.181$. The test statistic is

$$t = \frac{-.88}{1.181\sqrt{\frac{1}{14} + \frac{1}{12}}} = \frac{-.88}{.465} \approx -1.89 \text{ . The p-value} = 2[P(t_{24} > 1.9)] = 2(.035) = .070.$$

With the pooled method, there are more degrees of freedom, and the p-value is smaller than with the unpooled method. That is, if we are willing to assume equal variances (which might or might not be valid here), the pooled test is more capable of detecting a significant difference between the sample means.

83.

a. With n denoting the second sample size, the first is m = 3n. We then wish

$$20 = 2(2.58)\sqrt{\frac{900}{3n} + \frac{400}{n}}\,,$$ which yields n = 47, m = 141.

b. We wish to find the n which minimizes $2(z_{\alpha/2})\sqrt{\frac{900}{400-n} + \frac{400}{n}}$, or

equivalently, the n which minimizes $\frac{900}{400-n} + \frac{400}{n}$. Taking the derivative with

respect to *n* and equating to 0 yields $900(400-n)^{-2} - 400n^{-2} = 0$, whence

$9n^2 = 4(400-n)^2$, or $5n^2 + 3200n - 640{,}000 = 0$. This yields n = 160, m = 400

$-n = 240$.

85. We want to test the hypothesis $H_0: \mu_1 \le 1.5\mu_2$ v. $H_a: \mu_1 > 1.5\mu_2$ – or, using the hint, H_0:

$\theta \le 0$ v. $H_a: \theta > 0$. Our point estimate of θ is $\hat{\theta} = \overline{X}_1 - 1.5\overline{X}_2$, whose estimated

standard error equals $s(\hat{\theta}) = \sqrt{\frac{s_1^2}{n_1} + (1.5)^2\,\frac{s_2^2}{n_2}}$, using the fact that

$V(\hat{\theta}) = \frac{\sigma_1^2}{n_1} + (1.5)^2\,\frac{\sigma_2^2}{n_2}$. Plug in the values provided to get a test statistic $t =$

$\dfrac{22.63 - 1.5(14.15) - 0}{\sqrt{2.8975}} \approx 0.83$. A conservative df estimate here is $\nu = 50 - 1 = 49$, and

$t_{.05,49} \approx 1.676$. Since 0.83 < 1.676, we fail to reject H_0 at the 5% significance level.
The data does not suggest that the average tip after an introduction is more than 50%
greater than the average tip without introduction

87. $\Delta_0 = 0$, $\sigma_1 = \sigma_2 = 10$, $d = 1$, $\sigma = \sqrt{\dfrac{200}{n}} = \dfrac{14.142}{\sqrt{n}}$, so $\beta = \Phi\left(1.645 - \dfrac{\sqrt{n}}{14.142}\right)$,

giving $\beta = .9015, .8264, .0294$, and .0000 for $n = 25, 100, 2500$, and 10,000
respectively. If the μ_i's referred to true average IQ's resulting from two different
conditions, $\mu_1 - \mu_2 = 1$ would have little practical significance, yet very large sample
sizes would yield statistical significance in this situation.

89. $H_0 : p_1 = p_2$ will be rejected at level α in favor of $H_a : p_1 > p_2$ if either $z \geq z_{.05} = 1.645$. With $\hat{p}_1 = \frac{250}{2500} = .10$, $\hat{p}_2 = \frac{167}{2500} = .0668$, and $\hat{p} = .0834$,

$z = \dfrac{.0332}{.0079} = 4.2$, so H_0 is rejected. It appears that a response is more likely for a white name than for a black name.

91.

a. Let μ_1 and μ_2 denote the true average weights for operations 1 and 2, respectively. The relevant hypotheses are $H_0 : \mu_1 - \mu_2 = 0$ vs. $H_a : \mu_1 - \mu_2 \neq 0$. The value of the test statistic is

$t = \dfrac{(1402.24 - 1419.63)}{\sqrt{\dfrac{(10.97)^2}{30} + \dfrac{(9.96)^2}{30}}} = \dfrac{-17.39}{\sqrt{4.011363 + 3.30672}} = \dfrac{-17.39}{\sqrt{7.318083}} = -6.43$. The

d.f. $\nu = \dfrac{(7.318083)^2}{\dfrac{(4.011363)^2}{29} + \dfrac{(3.30672)^2}{29}} = 57.5$, so use df = 57. $t_{.025,57} \approx 2.000$, so

we can reject H_0 at level .05. The data indicates that there is a significant difference between the true mean weights of the packages for the two operations.

b. $H_0 : \mu_1 = 1400$ will be tested against $H_a : \mu_1 > 1400$ using a one-sample t test with test statistic $t = \dfrac{\bar{x} - 1400}{s_1 / \sqrt{m}}$. With degrees of freedom = 29, we reject H_0 if

$t > t_{.05,29} = 1.699$. The test statistic value is $t = \dfrac{1402.24 - 1400}{10.97 / \sqrt{30}} = \dfrac{2.24}{2.00} = 1.1$.

Because $1.1 < 1.699$, H_0 is not rejected. True average weight does not appear to exceed 1400.

93. A large-sample confidence interval for $\lambda_1 - \lambda_2$ is $(\hat{\lambda}_1 - \hat{\lambda}_2) \pm z_{\alpha/2} \sqrt{\dfrac{\hat{\lambda}_1}{m} + \dfrac{\hat{\lambda}_2}{n}}$, or

$(\bar{x} - \bar{y}) \pm z_{\alpha/2} \sqrt{\dfrac{\bar{x}}{m} + \dfrac{\bar{y}}{n}}$. With $\bar{x} = 1.62$ and $\bar{y} = 2.56$, the 95% confidence interval for $\lambda_1 - \lambda_2$ is $-.94 \pm 1.96(.177) = -.94 \pm .35 = (-1.29, -.59)$.

CHAPTER 10

Section 10.1

1.

a. H_0 will be rejected if $f \geq F_{.05,4,15} = 3.06$ (since $I - 1 = 4$, and $I(J - 1) = (5)(3) =$

   15). The computed value of F is $f = \dfrac{MSTr}{MSE} = \dfrac{2673.3}{1094.2} = 2.44$. Since 2.44 is not
   ≥ 3.06, H_0 is not rejected. The data does not indicate a difference in the mean
   tensile strengths of the different types of copper wires.

b. $F_{.05,4,15} = 3.06$ and $F_{.10,4,15} = 2.36$, and our computed value of 2.44 is between
   those values, so it can be said that $.05 < $ P-value $< .10$.

3. With μ_i = true average lumen output for brand i bulbs, we wish to test
   $H_0 : \mu_1 = \mu_2 = \mu_3$ versus H_a: at least two μ_i's are unequal.
   $MSTr = \hat{\sigma}_B^2 = \dfrac{591.2}{2} = 295.60$, $MSE = \hat{\sigma}_W^2 = \dfrac{4773.3}{21} = 227.30$, so
   $f = \dfrac{MSTr}{MSE} = \dfrac{295.60}{227.30} = 1.30$ For finding the P-value, we need degrees of freedom $I - 1 = 2$ and $I(J - 1) = 21$. In the 2nd row and 21st column of Table A.9, we see that
   $1.30 < F_{.10,2,21} = 2.57$, so the P-value $> .10$. Since .10 is not $< .05$, we cannot reject
   H_0. There are no differences in the average lumen outputs among the three brands of
   bulbs.

5. μ_i = true mean modulus of elasticity for grade i ($i = 1, 2, 3$). We test
   $H_0 : \mu_1 = \mu_2 = \mu_3$ vs. H_a: at least two μ_i's are unequal. Reject H_0 if
   $f \geq F_{.01,2,27} = 5.49$. The grand mean = 1.5367,
   $$MSTr = \dfrac{10}{2}\left[(1.63 - 1.5367)^2 + (1.56 - 1.5367)^2 + (1.42 - 1.5367)^2\right] = .1143,$$
   $$MSE = \dfrac{1}{3}\left[(.27)^2 + (.24)^2 + (.26)^2\right] = .0660, \ f = \dfrac{MSTr}{MSE} = \dfrac{.1143}{.0660} = 1.73.$$ Hence, we
   fail to reject H_0. The three grades do not appear to differ significantly.

7.

Source	df	SS	MS	F
Treatments	3	75,081.72	25,027.24	1.70
Error	16	235,419.04	14,713.69	
Total	19	310,500.76		

The hypotheses are $H_0 : \mu_1 = \mu_2 = \mu_3 = \mu_4$ vs. H_a : at least two μ_i's are unequal. $1.70 < F_{.10,3,16} = 2.46$, so P-value $> .10$, and we fail to reject H_0.

9.

The summary quantities are $x_{1\bullet} = 34.3$, $x_{2\bullet} = 39.6$, $x_{3\bullet} = 33.0$, $x_{4\bullet} = 41.9$,

$x_{\bullet\bullet} = 148.8$, $\Sigma\Sigma x_{ij}^2 = 946.68$, so $CF = \dfrac{(148.8)^2}{24} = 922.56$,

$SST = 946.68 - 922.56 = 24.12$, $SSTr = \dfrac{(34.3)^2 + ... + (41.9)^2}{6} - 922.56 = 8.98$,

$SSE = 24.12 - 8.98 = 15.14$.

Source	df	SS	MS	F
Treatments	3	8.98	2.99	3.95
Error	20	15.14	.757	
Total	23	24.12		

Since $3.10 = F_{.05,3,20} < 3.95 < 4.94 = F_{.01,3,20}$, $.01 <$ P-value $< .05$ and H_0 is rejected at level $.05$.

Section 10.2

11. $Q_{.05,5,15} = 4.37$, $w = 4.37\sqrt{\dfrac{272.8}{4}} = 36.09$.

3	1	4	2	5
437.5	462.0	469.3	512.8	532.1

The brands seem to divide into two groups: 1, 3, and 4; and 2 and 5; with no significant differences within each group but all between group differences are significant.

13.

3	1	4	2	5
427.5	462.0	469.3	502.8	532.1

Brand 1 does not differ significantly from 3 or 4, 2 does not differ significantly from 4 or 5, 3 does not differ significantly from1, 4 does not differ significantly from 1 or 2, 5 does not differ significantly from 2, but all other differences (e.g., 1 with 2 and 5, 2 with 3, etc.) do appear to be significant.

15. $Q_{.01,4,36} = 4.75$, $w = 4.75\sqrt{\dfrac{15.64}{10}} = 5.94$.

2	1	3	4
24.69	26.08	29.95	33.84

Treatment 4 appears to differ significantly from both 1 and 2, but there are no other significant differences.

Chapter 10: The Analysis of Variance

17. $\theta = \Sigma c_i \mu_i$ where $c_1 = c_2 = .5$ and $c_3 = -1$, so $\hat{\theta} = .5\bar{x}_{1\bullet} + .5\bar{x}_{2\bullet} - \bar{x}_{3\bullet} = -.527$ and $\Sigma c_i^2 = 1.50$. With $t_{.025,27} = 2.052$ and MSE $= .0660$, the desired CI is

$$-.527 \pm (2.052)\sqrt{\frac{(.0660)(1.50)}{10}} = -.527 \pm .204 = (-.731, -.323).$$

19. MSTr $= 140$, error df $= 12$, so $f = \dfrac{140}{SSE/12} = \dfrac{1680}{SSE}$ and $F_{.05,2,12} = 3.89$.

$w = Q_{.05,3,12}\sqrt{\dfrac{MSE}{J}} = 3.77\sqrt{\dfrac{SSE}{60}} = .4867\sqrt{SSE}$. Thus we wish $\dfrac{1680}{SSE} > 3.89$

(significance of f) and $.4867\sqrt{SSE} > 10$ ($= 20 - 10$, the difference between the extreme $\bar{x}_{i\bullet}$'s $-$ so no significant differences are identified). These become $431.88 > SSE$ and $SSE > 422.16$, so SSE $= 425$ will work.

21.

a. Grand mean $= 222.167$, MSTr $= 38,015.1333$, MSE $= 1,681.8333$, and $f = 22.6$. The hypotheses are $H_0 : \mu_1 = ... = \mu_6$ vs. H_a : at least two μ_i's differ. Reject H_0 if $f \geq F_{.01,5,78}$ (but since there is no table value for $\nu_2 = 78$, use $f \geq F_{.01,5,60} = 3.34$) With $22.6 \geq 3.34$, we reject H_0. The data indicates there is a dependence on injection regimen.

b. Assume $t_{.005,78} \approx 2.645$

i) Confidence interval for $\mu_1 - \frac{1}{5}(\mu_2 + \mu_3 + \mu_4 + \mu_5 + \mu_6)$:

$$\Sigma c_i \bar{x}_i \pm t_{\alpha/2, I(J-1)}\sqrt{\frac{MSE(\Sigma c_i^2)}{J}}$$

$$= -67.4 \pm (2.645)\sqrt{\frac{1,681.8333(1.2)}{14}} = (-99.16, -35.64).$$

ii) Confidence interval for $\frac{1}{4}(\mu_2 + \mu_3 + \mu_4 + \mu_5) - \mu_6$:

$$= 61.75 \pm (2.645)\sqrt{\frac{1,681.8333(1.25)}{14}} = (29.34, 94.16)$$

Section 10.3

23. $J_1 = 5$, $J_2 = 4$, $J_3 = 4$, $J_4 = 5$, $\bar{x}_{1\bullet} = 58.28$, $\bar{x}_{2\bullet} = 55.40$, $\bar{x}_{3\bullet} = 50.85$, $\bar{x}_{4\bullet} = 45.50$,

$MSE = 8.89$. With $W_{ij} = Q_{.05,4,14} \cdot \sqrt{\dfrac{MSE}{2}\left(\dfrac{1}{J_i}+\dfrac{1}{J_j}\right)} = 4.11\sqrt{\dfrac{8.89}{2}\left(\dfrac{1}{J_i}+\dfrac{1}{J_j}\right)}$,

$\bar{x}_{1\bullet} - \bar{x}_{2\bullet} \pm W_{12} = (2.88)\pm(5.81)$; $\quad \bar{x}_{1\bullet} - \bar{x}_{3\bullet} \pm W_{13} = (7.43)\pm(5.81)*$;

$\bar{x}_{1\bullet} - \bar{x}_{4\bullet} \pm W_{14} = (12.78)\pm(5.48)*$; $\bar{x}_{2\bullet} - \bar{x}_{3\bullet} \pm W_{23} = (4.55)\pm(6.13)$;

$\bar{x}_{2\bullet} - \bar{x}_{4\bullet} \pm W_{24} = (9.90)\pm(5.81)*$; $\bar{x}_{3\bullet} - \bar{x}_{4\bullet} \pm W_{34} = (5.35)\pm(5.81)$;

*Indicates an interval that doesn't include zero, corresponding to μ's that are judged significantly different.

$$
\begin{array}{cccc}
4 & 3 & 2 & 1 \\
\underline{} & & & \\
& & \underline{} &
\end{array}
$$

This underscoring pattern does not have a very straightforward interpretation.

25.

a. The distributions of the polyunsaturated fat percentages for each of the four regimens must be normal with equal variances.

b. We have all the $\bar{X}_{i\bullet}$'s, and we need the grand mean:

$\bar{X}_{\bullet\bullet} = \dfrac{8(43.0)+13(42.4)+17(43.1)+14(43.5)}{52} = \dfrac{2236.9}{52} = 43.017$;

$SSTr = \sum J_i(\bar{x}_{i\bullet} - \bar{x}_{\bullet\bullet})^2 = 8(43.0 - 43.017)^2 + 13(42.4 - 43.017)^2$

$+ 17(43.1 - 43.017)^2 + 13(43.5 - 43.017)^2 = 8.334$ and $MSTr = \dfrac{8.334}{3} = 2.778$

$SSE = \sum (J_i - 1)s_i^2 = 7(1.5)^2 + 12(1.3)^2 + 16(1.2)^2 + 13(1.2)^2 = 77.79$ and

$MSE = \dfrac{77.79}{48} = 1.621$. Then $f = \dfrac{MSTr}{MSE} = \dfrac{2.778}{1.621} = 1.714$

Since $1.714 < F_{.10,3,50} = 2.20$, we can say that the P-value is $> .10$. We do not reject the null hypothesis at significance level .10 (or any smaller), so we conclude that the data suggests no difference in the percentages for the different regimens.

27.

a. Let μ_i = true average folacin content for specimens of brand I. The hypotheses to be tested are $H_0 : \mu_1 = \mu_2 = \mu_3 = \mu_4$ vs. H_a : at least two μ_i's differ .

$$\Sigma\Sigma x_{ij}^2 = 1246.88 \text{ and } \frac{x_{\bullet\bullet}^2}{n} = \frac{(168.4)^2}{24} = 1181.61 \text{, so SST} = 65.27.$$

$$\frac{\Sigma x_{i\bullet}^2}{J_i} = \frac{(57.9)^2}{7} + \frac{(37.5)^2}{5} + \frac{(38.1)^2}{6} + \frac{(34.9)^2}{6} = 1205.10 \text{, so}$$

$$SSTr = 1205.10 - 1181.61 = 23.49.$$

Source	df	SS	MS	F
Treatments	3	23.49	7.83	3.75
Error	20	41.78	2.09	
Total	23	65.27		

With num df = 3 and den df = 20, $F_{.05,3,20} = 3.10 < 3.75 < F_{.01,3,20} = 4.94$, so .01 < P-value < .05, and since the P-value < .05, we reject H_0. At least one of the pairs of brands of green tea has different average folacin content.

b. With $\bar{x}_{i\bullet} = 8.27, 7.50, 6.35$, and 5.82 for $i = 1, 2, 3, 4$, we calculate the residuals $x_{ij} - \bar{x}_{i\bullet}$ for all observations. A normal probability plot appears below, and indicates that the distribution of residuals could be normal, so the normality assumption is plausible. The sample standard deviations are 1.463, 1.681, 1.060, and 1.551, so the equal variance assumption is plausible (since the largest sd is less than twice the smallest sd).

Normal Probability Plot for ANOVA Residuals

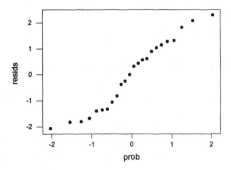

c. $Q_{.05,4,20} = 3.96$ and $W_{ij} = 3.96 \cdot \sqrt{\dfrac{2.09}{2}\left(\dfrac{1}{J_i} + \dfrac{1}{J_j}\right)}$. The Modified Tukey

intervals are as follows. Only Brands 1 and 4 are different from each other.

Pair	Interval	Pair	Interval
1,2	.77 ± 2.37	2,3	1.15 ± 2.45
1,3	1.92 ± 2.25	2,4	1.68 ± 2.45
1,4	2.45 ± 2.25 *	3,4	.53 ± 2.34

$$4 \qquad 3 \qquad 2 \qquad 1$$

29. $E(SSTr) = E\left(\sum_i J_i \overline{X}_{i\bullet}^2 - n\overline{X}_{\bullet\bullet}^2\right) = \sum J_i E\left(\overline{X}_{i\bullet}^2\right) - nE\left(\overline{X}_{\bullet\bullet}^2\right)$

$= \sum J_i\left[Var\left(\overline{X}_{i\bullet}\right) + \left(E\left(\overline{X}_{i\bullet}\right)\right)^2\right] - n\left[Var\left(\overline{X}_{\bullet\bullet}\right) + \left(E\left(\overline{X}_{\bullet\bullet}\right)\right)^2\right]$

$= \sum J_i\left[\dfrac{\sigma^2}{J_i} + \mu_i^2\right] - n\left[\dfrac{\sigma^2}{n} + \left(\dfrac{\sum J_i \mu_i}{n}\right)^2\right]$

$= I\sigma^2 + \sum J_i(\mu + \alpha_i)^2 - \sigma^2 - \dfrac{1}{n}\left[\sum J_i(\mu + \alpha_i)\right]^2$

$= (I-1)\sigma^2 + \sum J_i \mu^2 + 2\mu \sum J_i \alpha_i + \sum J_i \alpha_i^2 - \dfrac{1}{n}\left[n\mu + 0\right]^2$

$= (I-1)\sigma^2 + \mu^2 n + 2\mu 0 + \sum J_i \alpha_i^2 - n\mu^2 = (I-1)\sigma^2 + \sum J_i \alpha_i^2$, from which E(MSTr)
is obtained through division by $(I-1)$.

31. With $\sigma = 1$ (any other σ would yield the same Φ), $\alpha_1 = -1$, $\alpha_2 = \alpha_3 = 0$, $\alpha_4 = 1$,

$\Phi^2 = \dfrac{1\left(5(-1)^2 + 4(0)^2 + 4(0)^2 + 5(1)^2\right)}{4} = 2.5$, $\Phi = 1.58$, $v_1 = 3$, $v_2 = 14$, and power

$\approx .65$.

33. $g(x) = x\left(1 - \dfrac{x}{n}\right) = nu(1-u)$ where $u = \dfrac{x}{n}$, so $h(x) = \int\left[u(1-u)\right]^{-1/2} du$. This gives

$h(x) = \arcsin\left(\sqrt{u}\right) = \arcsin\left(\sqrt{\dfrac{x}{n}}\right)$ as the appropriate transformation.

Chapter 10: The Analysis of Variance

Supplementary Exercises

35.

a. $H_0 : \mu_1 = \mu_2 = \mu_3 = \mu_4$ vs. H_a : at least two $\mu_i 's$ differ ; 3.68 is not $\geq F_{.01,3,20} = 4.94$, thus fail to reject H_0. The means do not appear to differ.

b. We reject H_0 when the P-value $< \alpha$. Since .029 $>$ .01, we still fail to reject H_0.

37. Let μ_i = true average amount of motor vibration for each of five bearing brands. Then the hypotheses are $H_0 : \mu_1 = ... = \mu_5$ vs. H_a : at least two $\mu_i 's$ differ. The ANOVA table follows:

Source	df	SS	MS	F
Treatments	4	30.855	7.714	8.44
Error	25	22.838	0.914	
Total	29	53.694		

$8.44 > F_{.001,4,25} = 6.49$, so P-value $< .001 < .05$, and we reject H_0. At least two of the means differ from one another. The Tukey multiple comparison procedure is appropriate. $Q_{.05,5,25} = 4.15$ (from Minitab output; using Table A.10, we approximate with $Q_{.05,5,24} = 4.17$). $W_{ij} = 4.15\sqrt{.914/6} = 1.620$.

Pair	$\bar{x}_{i\bullet} - \bar{x}_{j\bullet}$	Pair	$\bar{x}_{i\bullet} - \bar{x}_{j\bullet}$
1,2	−2.267*	2,4	1.217
1,3	0.016	2,5	2.867*
1,4	−1.050	3,4	−1.066
1,5	0.600	3,5	0.584
2,3	2.283*	4,5	1.650*

*Indicates significant pairs.

5	3	1	4	2

39. $\hat{\theta} = 2.58 - \dfrac{2.63 + 2.13 + 2.41 + 2.49}{4} = .165$, $t_{.025,25} = 2.060$, MSE = .108, and

$\Sigma c_i^2 = (1)^2 + (-.25)^2 + (-.25)^2 + (-.25)^2 + (-.25)^2 = 1.25$, so a 95% confidence

interval for θ is $.165 \pm 2.060 \sqrt{\dfrac{(.108)(1.25)}{6}} = .165 \pm .309 = (-.144, .474)$. This interval

does include zero, so 0 is a plausible value for θ.

41. This is a random effects problem. $H_0 : \sigma_A^2 = 0$ states that variation in laboratories doesn't contribute to variation in percentage. H_0 will be rejected in favor of H_a if $f \geq F_{.05,3,8} = 4.07$. SST = 86,078.9897 − 86,077.2224 = 1.7673, SSTr = 1.0559, and

SSE = .7114. Thus $f = \dfrac{1.0559/3}{0.7114/8} = 3.96$, which is not ≥ 4.07, so H_0 cannot be

rejected at level .05. Variation in laboratories does not appear to be present.

43. $\sqrt{(I-1)(MSE)(F_{.05,I-1,n-I})} = \sqrt{(2)(2.39)(3.63)} = 4.166$. For $\mu_1 - \mu_2$, $c_1 = 1$, $c_2 = -1$,

and $c_3 = 0$, so $\sqrt{\Sigma \dfrac{c_i^2}{J_i}} = \sqrt{\dfrac{1}{8} + \dfrac{1}{5}} = .570$. Similarly, for $\mu_1 - \mu_3$,

$\sqrt{\Sigma \dfrac{c_i^2}{J_i}} = \sqrt{\dfrac{1}{8} + \dfrac{1}{6}} = .540$; for $\mu_2 - \mu_3$, $\sqrt{\Sigma \dfrac{c_i^2}{J_i}} = \sqrt{\dfrac{1}{5} + \dfrac{1}{6}} = .606$, and for

$.5\mu_2 + .5\mu_2 - \mu_3$, $\sqrt{\Sigma \dfrac{c_i^2}{J_i}} = \sqrt{\dfrac{.5^2}{8} + \dfrac{.5^2}{5} + \dfrac{(-1)^2}{6}} = .498$.

Contrast	Estimate	Interval
$\mu_1 - \mu_2$	$25.59 - 26.92 = -1.33$	$(-1.33) \pm (.570)(4.166) = (-3.70, 1.04)$
$\mu_1 - \mu_3$	$25.59 - 28.17 = -2.58$	$(-2.58) \pm (.540)(4.166) = (-4.83, -.33)$
$\mu_2 - \mu_3$	$26.92 - 28.17 = -1.25$	$(-1.25) \pm (.606)(4.166) = (-3.77, 1.27)$
$.5\mu_2 + .5\mu_2 - \mu_3$	-1.92	$(-1.92) \pm (.498)(4.166) = (-3.99, 0.15)$

45. $Y_{ij} - \overline{Y}_{\bullet\bullet} = c\left(X_{ij} - \overline{X}_{\bullet\bullet}\right)$ and $\overline{Y}_{i\bullet} - \overline{Y}_{\bullet\bullet} = c\left(\overline{X}_{i\bullet} - \overline{X}_{\bullet\bullet}\right)$, so each sum of squares involving Y will be the corresponding sum of squares involving X multiplied by c^2. Since F is a ratio of two sums of squares, c^2 appears in both the numerator and denominator so cancels, and F computed from Y_{ij}'s $= F$ computed from X_{ij}'s.

CHAPTER 11

Section 11.1

1.

a. $MSA = \dfrac{30.6}{4} = 7.65$, $MSE = \dfrac{59.2}{12} = 4.93$, $f_A = \dfrac{7.65}{4.93} = 1.55$. Since 1.55 is not $\geq F_{.05,4,12} = 3.26$, don't reject H_{0A}. There is no significant difference in true average tire lifetime due to different makes of cars.

b. $MSB = \dfrac{44.1}{3} = 14.70$, $f_B = \dfrac{14.70}{4.93} = 2.98$. Since 2.98 is not $\geq F_{.05,3,12} = 3.49$, don't reject H_{0B}. There is no significant difference in true average tire lifetime due to different brands of tires.

3. Software provides the ANOVA table below. Since $F_{.01,3,9} = 6.99$, both H_{0A} and H_{0B} are rejected.

a.

Source	DF	SS	MS	F
A	3	324,082.2	108,027.4	105.3
B	3	39,934.2	13,311.4	13.0
Error	9	9232.0	1025.8	
Total	15	373,248.4		

b. $Q_{.01,4,9} = 5.96$, $w = 5.96\sqrt{\dfrac{1025.8}{4}} = 95.4$

i:	1	2	3	4
$\bar{x}_{i\bullet}$:	231.75	325.25	441.00	613.25

All levels of Factor A (gas rate) differ significantly except for 1 and 2.

c. $w = 95.4$, as in b

j:	1	2	3	4
$\bar{x}_{\bullet j}$:	336.75	382.25	419.25	473

Only levels 1 and 4 appear to differ significantly.

Chapter 11: Multifactor Analysis of Variance

5.

Source	DF	SS	MS	F
Angle	3	58.16	19.3867	2.5565
Connector	4	246.97	61.7425	8.1419
Error	12	91.00	7.5833	
Total	19	396.13		

$H_0 : \alpha_1 = \alpha_2 = \alpha_3 = \alpha_4 = 0$; H_a : at least one α_i is not zero.
$f_A = 2.5565 < F_{.01,3,12} = 5.95$, so fail to reject H_0. The data fails to indicate any effect due to the angle of pull.

7.

a. SSE = SST – SSTr – SSBl = 3476.00 – 28.78 – 2977.67 = 469.55; MSTr = 28.78/(3–1) = 14.39, MSE = SSE/[(18–1)(3–1)] = 13.81. Hence, f_{Tr} = 1.04, which is clearly insignificant when compared to $F_{.05,2,34}$.

b. $f_{Bl} = 12.68$, which is significant, and suggests substantial variation among subjects. If we had not controlled for such variation, it might have affected the analysis and conclusions.

9.

Source	Df	SS	MS	f
Treatment	3	81.1944	27.0648	22.36
Block	8	66.5000	8.3125	6.87
Error	24	29.0556	1.2106	
Total	35	176.7500		

$F_{.05,3,24} = 3.01$. Reject H_0. There is an effect due to treatments.

$Q_{.05,4,24} = 3.90$; $w = (3.90)\sqrt{\dfrac{1.2106}{9}} = 1.43$

1	4	3	2
8.56	9.22	10.78	12.44

11. The residual, percentile pairs are (–0.1225, –1.73), (–0.0992, –1.15), (–0.0825, –0.81), (–0.0758, –0.55), (–0.0750, –0.32), (0.0117, –0.10), (0.0283, 0.10), (0.0350, 0.32), (0.0642, 0.55), (0.0708, 0.81), (0.0875, 1.15), (0.1575, 1.73). The pattern is sufficiently linear, so normality is plausible.

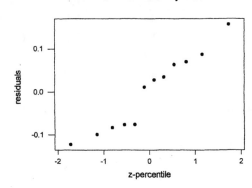

13.

a. With $Y_{ij} = X_{ij} + d$, $\overline{Y}_{i\bullet} = \overline{X}_{i\bullet} + d$, $\overline{Y}_{\bullet j} = \overline{X}_{\bullet j} + d$, $\overline{Y}_{\bullet\bullet} = \overline{X}_{\bullet\bullet} + d$, so all quantities inside the parentheses in (11.5) remain unchanged when the Y quantities are substituted for the corresponding X's (e.g., $\overline{Y}_{i\bullet} - \overline{Y}_{\bullet\bullet} = \overline{X}_{i\bullet} - \overline{X}_{\bullet\bullet}$, etc.).

b. With $Y_{ij} = cX_{ij}$, each sum of squares for Y is the corresponding SS for X multiplied by c^2. However, when F ratios are formed the c^2 factors cancel, so all F ratios computed from Y are identical to those computed from X. If $Y_{ij} = cX_{ij} + d$, the conclusions reached from using the Y's will be identical to those reached using the X's.

15.

a. $\Sigma\alpha_i^2 = 24$, so $\Phi^2 = \left(\dfrac{3}{4}\right)\left(\dfrac{24}{16}\right) = 1.125$, $\Phi = 1.06$, $\nu_1 = 3$, $\nu_2 = 6$, and from figure 10.5, power $\approx .2$. For the second alternative, $\Phi = 1.59$, and power $\approx .43$.

b. $\Phi^2 = \left(\dfrac{I}{J}\right)\Sigma\dfrac{\beta_j^2}{\sigma^2} = \left(\dfrac{4}{5}\right)\left(\dfrac{20}{16}\right) = 1.00$, so $\Phi = 1.00$, $\nu_1 = 4$, $\nu_2 = 12$, power $\approx .3$.

Section 11.2

17.

a.

Source	df	SS	MS	f	$F_{.05}$
Sand	2	705	352.5	3.76	4.26
Fiber	2	1,278	639.0	6.82*	4.26
Sand&Fiber	4	279	69.75	0.74	3.63
Error	9	843	93.67		
Total	17	3,105			

There appears to be an effect due to carbon fiber addition.

b.

Source	df	SS	MS	f	$F_{.05}$
Sand	2	106.78	53.39	6.54*	4.26
Fiber	2	87.11	43.56	5.33*	4.26
Sand&Fiber	4	8.89	2.22	.27	3.63
Error	9	73.50	8.17		
Total	17	276.28			

There appears to be an effect due to both sand and carbon fiber addition to casting hardness.

c.

Sand%	0	15	30	0	15	30	0	15	30
Fiber%	0	0	0	0.25	0.25	0.25	0.5	0.5	0.5
$\bar{x}$	62	68	69.5	69	71.5	73	68	71.5	74

The interaction plot on the next page indicates some effect due to sand and fiber addition with no significant interaction. This agrees with the statistical analysis in part b.

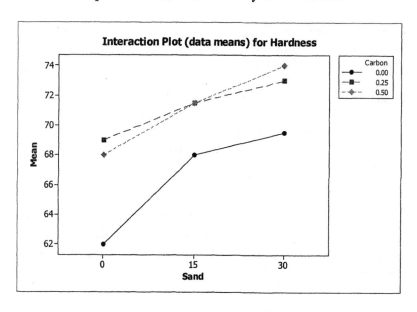

19.

a. Software gives the ANOVA table below. Call coal type Factor A and NaOH concentration Factor B. H_{0AB} cannot be rejected ($f = 0.21$, P-value = .924), so no significant interaction is present. H_{0B} cannot be rejected ($f = 3.66$, P-value = .069), so varying levels of NaOH does not have a significant impact on total acidity. H_{0A} is rejected at $\alpha = .01$ ($f = 29.49$, P-value = .000): type of coal does appear to affect total acidity.

Two-way ANOVA: Acidity versus Coal, NaOH

Source	DF	SS	MS	F	P
Coal	2	1.00241	0.501206	29.49	0.000
NaOH	2	0.12431	0.062156	3.66	0.069
Interaction	4	0.01456	0.003639	0.21	0.924
Error	9	0.15295	0.016994		
Total	17	1.29423			

b. $Q_{.01,3,9} = 5.43$, $w = 5.43\sqrt{\dfrac{.0170}{6}} = .289$

i:	3	1	2
$\bar{x}_{i..}$	8.035	8.247	8.607

Coal 2 is judged significantly different from both 1 and 3, but these latter two don't differ significantly from each other.

21. From the provided sums of squares,
$SSAB = 64,954.70 - [22,941.80 + 22,765.53 + 15,253.50] = 3993.87$. This allows us to complete the ANOVA table below.

Source	df	SS	MS	f
A	2	22,941.80	11,470.90	22.98
B	4	22,765.53	5691.38	11.40
AB	8	3993.87	499.23	.49
Error	15	15,253.50	1016.90	
Total	29	64,954.70		

$f_{AB} = .49$ is clearly not significant. Since $22.98 \geq F_{.05,2,8} = 4.46$, H_{0A} is rejected; since $11.40 \geq F_{.05,4,8} = 3.84$, H_{0B} is also rejected. We conclude that the different cement factors affect flexural strength differently and that batch variability contributes to variation in flexural strength.

23. Summary quantities include $x_{1\bullet\bullet} = 9410$, $x_{2\bullet\bullet} = 8835$, $x_{3\bullet\bullet} = 9234$, $x_{\bullet1\bullet} = 5432$, $x_{\bullet2\bullet} = 5684$, $x_{\bullet3\bullet} = 5619$, $x_{\bullet4\bullet} = 5567$, $x_{\bullet3\bullet} = 5177$, $x_{\bullet\bullet\bullet} = 27,479$, $CF = 16,779,898.69$, $\Sigma x_{i\bullet\bullet}^2 = 251,872,081$, $\Sigma x_{\bullet j\bullet}^2 = 151,180,459$, resulting in the accompanying ANOVA table.

Source	Df	SS	MS	f
A	2	11,573.38	5786.69	$\frac{MSA}{MSAB} = 26.70$
B	4	17,930.09	4482.52	$\frac{MSB}{MSAB} = 20.68$
AB	8	1734.17	216.77	$\frac{MSAB}{MSE} = 1.38$
Error	30	4716.67	157.22	
Total	44	35,954.31		

Since $1.38 < F_{.01,8,30} = 3.17$, H_{0G} cannot be rejected, and we continue:
$26.70 \geq F_{.01,2,8} = 8.65$, and $20.68 \geq F_{.01,4,8} = 7.01$, so both H_{0A} and H_{0B} are rejected. Both capping material and the different batches affect compressive strength of concrete cylinders.

25. With $\theta = \alpha_i - \alpha_i'$, $\hat\theta = \overline{X}_{i..} - \overline{X}_{i'..} = \dfrac{1}{JK}\sum_j \sum_k \left(X_{ijk} - X_{i'jk}\right)$, and since $i \neq i'$,

X_{ijk} and $X_{i'jk}$ are independent for every j, k. Thus,

$$Var(\hat\theta) = Var(\overline{X}_{i..}) + Var(\overline{X}_{i'..}) = \frac{\sigma^2}{JK} + \frac{\sigma^2}{JK} = \frac{2\sigma^2}{JK} \text{ (because } Var(\overline{X}_{i..}) = Var(\overline{\varepsilon}_{i..}) \text{ and }$$

$Var(\varepsilon_{ijk}) = \sigma^2$) so $\hat\sigma_{\hat\theta} = \sqrt{\dfrac{2MSE}{JK}}$. The appropriate number of df is $IJ(K-1)$, so the

C.I. is $\left(\overline{x}_{i..} - \overline{x}_{i'..}\right) \pm t_{\alpha/2, IJ(K-1)}\sqrt{\dfrac{2MSE}{JK}}$. For the data of exercise 19, $\overline{x}_{2..} = 49.15$,

$\overline{x}_{3..} = 50.37$, MSE $= .0170$, $t_{.025,9} = 2.262$, J $= 3$, K $= 2$, so the C.I. is

$\left(49.15 - 50.37\right) \pm 2.262\sqrt{\dfrac{.0340}{6}} = -1.22 \pm .17 = \left(-1.39, -1.05\right)$.

Section 11.3

27.

a.

Source	Df	SS	MS	f	$F_{.05}$
A	2	14,144.44	7072.22	61.06	3.35
B	2	5,511.27	2755.64	23.79	3.35
C	2	244,696.39	122.348.20	1056.24	3.35
AB	4	1,069.62	267.41	2.31	2.73
AC	4	62.67	15.67	.14	2.73
BC	4	331.67	82.92	.72	2.73
ABC	8	1,080.77	135.10	1.17	2.31
Error	27	3,127.50	115.83		
Total	53	270,024.33			

b. The computed F-statistics for all four interaction terms are less than the tabled values for statistical significance at the level .05. This indicates that none of the interactions are statistically significant.

c. The computed F-statistics for all three main effects exceed the tabled value for significance at level .05. All three main effects are statistically significant.

d. Since $Q_{.05,3,27}$ is not tabled, use $Q_{.05,3,24} = 3.53$, $w = 3.53\sqrt{\dfrac{115.83}{(3)(3)(2)}} = 8.95$. All three levels differ significantly from each other.

29. $I = 3, J = 2, K = 4, L = 4;\ \bar{x}_{1\cdots} = 3.781;\ SSB = IKL\sum\left(\bar{x}_{\cdot j\cdot} - \bar{x}_{\cdots}\right)^2;$

$SSC = IJL\sum\left(\bar{x}_{\cdot\cdot k\cdot} - \bar{x}_{\cdots}\right)^2.$

For level A: $\bar{x}_{1\cdots} = 3.781$ $\bar{x}_{2\cdots} = 3.625$ $\bar{x}_{3\cdots} = 4.469$

For level B: $\bar{x}_{\cdot 1\cdot} = 4.979$ $\bar{x}_{\cdot 2\cdot} = 2.938$

For level C: $\bar{x}_{\cdot\cdot 1} = 3.417$ $\bar{x}_{\cdot\cdot 2} = 5.875$ $\bar{x}_{\cdot\cdot 3} = .875$ $\bar{x}_{\cdot\cdot 4} = 5.667$

$\bar{x}_{\cdots} = 3.958$

SSA = 12.907;   SSB = 99.976;   SSC = 393.436

a.

Source	Df	SS	MS	f	$F_{.05}$*
A	2	12.907	6.454	1.04	3.15
B	1	99.976	99.976	16.09	4.00
C	3	393.436	131.145	21.10	2.76
AB	2	1.646	.823	.13	3.15
AC	6	71.021	11.837	1.90	2.25
BC	3	1.542	.514	.08	2.76
ABC	6	9.805	1.634	.26	2.25
Error	72	447.500	6.215		
Total	95	1,037.833			

*use 60 df for denominator of tabled F.

b.   No interaction effects are significant at level .05.

c.   Factor B and C main effects are significant at the level .05.

d. $Q_{.05,4,72}$ is not tabled, use $Q_{.05,4,60} = 3.74$, $w = 3.74\sqrt{\dfrac{6.215}{(3)(2)(4)}} = 1.90$.

Machine:	3	1	4	2
Mean:	.875	3.417	5.667	5.875

31.

$x_{ij.}$	B_1	B_2	B_3
A_1	210.2	224.9	218.1
A_2	224.1	229.5	221.5
A_3	217.7	230.0	202.0
$x_{.j.}$	652.0	684.4	641.6

$x_{i.k}$	A_1	A_2	A_3
C_1	213.8	222.0	205.0
C_2	225.6	226.5	223.5
C_3	213.8	226.6	221.2
$x_{i..}$	653.2	675.1	649.7

$x_{.jk}$	C_1	C_2	C_3
B_1	213.5	220.5	218.0
B_2	214.3	246.1	224.0
B_3	213.0	209.0	219.6
$x_{..k}$	640.8	675.6	661.6

$\Sigma\Sigma x_{ij.}^2 = 435,382.26$ $\Sigma\Sigma x_{i.k}^2 = 435,156.74$ $\Sigma\Sigma x_{.jk}^2 = 435,666.36$

$\Sigma x_{.j.}^2 = 1,305,157.92$ $\Sigma x_{i..}^2 = 1,304,540.34$ $\Sigma x_{..k}^2 = 1,304,774.56$

Also, $\Sigma\Sigma\Sigma x_{ijk}^2 = 145,386.40$, $x_{...} = 1978$, CF = 144,906.81, from which we obtain the ANOVA table displayed in the problem statement. $F_{.01,4,8} = 7.01$, so the AB and BC interactions are significant (as can be seen from the P-values) and tests for main effects are not appropriate.

33.

Source	df	SS	MS	f
A	6	67.32	11.02	
B	6	51.06	8.51	
C	6	5.43	.91	.61
Error	30	44.26	1.48	
Total	48	168.07		

Since $.61 < F_{.05,6,30} = 2.42$, treatment was not effective.

35.

	1	2	3	4	5	
$x_{i..}$	40.68	30.04	44.02	32.14	33.21	$\Sigma x_{i..}^2 = 6630.91$
$x_{.j.}$	29.19	31.61	37.31	40.16	41.82	$\Sigma x_{.j.}^2 = 6605.02$
$x_{..k}$	36.59	36.67	36.03	34.50	36.30	$\Sigma x_{..k}^2 = 6489.92$

$x_{...} = 180.09$, CF = 1297.30, $\Sigma\Sigma x_{ij(k)}^2 = 1358.60$

Source	df	SS	MS	f
A	4	28.89	7.22	10.78
B	4	23.71	5.93	8.85
C	4	0.69	0.17	0.25
Error	12	8.01	.67	
Total	24	61.30		

$F_{.05,4,12} = 3.26$, so both factor A (plant) and B(leaf size) appear to affect moisture content, but factor C (time of weighing) does not.

37.

Source	Df	MS	f	$F_{.01}$*
A	2	2207.329	2259.29	5.39
B	1	47.255	48.37	7.56
C	2	491.783	503.36	5.39
D	1	.044	.05	7.56
AB	2	15.303	15.66	5.39
AC	4	275.446	281.93	4.02
AD	2	.470	.48	5.39
BC	2	2.141	2.19	5.39
BD	1	.273	.28	7.56
CD	2	.247	.25	5.39
ABC	4	3.714	3.80	4.02
ABD	2	4.072	4.17	5.39
ACD	4	.767	.79	4.02
BCD	2	.280	.29	5.39
ABCD	4	.347	.355	4.02
Error	36	.977		
Total	71			

*Because denominator df for 36 is not tabled, use df = 30

SST = (71)(93.621) = 6,647.091. Computing all other sums of squares and adding them up = 6,645.702. Thus SSABCD = 6,647.091 − 6,645.702 = 1.389 and

$$MSABCD = \frac{1.389}{4} = .347 \,.$$

At level .01 the statistically significant main effects are A, B, C. The interaction AB and AC are also statistically significant. No other interactions are statistically significant.

Section 11.4

39.

Condition	Total	1	2	Contrast	$SS = \frac{(contrast)^2}{24}$
111	315	927	2478	5485	
211	612	1551	3007	1307	A = 71,177.04
121	584	1163	680	1305	B = 70,959.38
221	967	1844	627	199	AB = 1650.04
112	453	297	624	529	C = 11,660.04
212	710	383	681	−53	AC = 117.04
122	737	257	86	57	BC = 135.38
222	1107	370	113	27	ABC = 30.38

a. $\hat{\beta}_1 = \bar{x}_{.2..} - \bar{x}_{....} = \dfrac{584 + 967 + 737 + 1107 - 315 - 612 - 453 - 710}{24} = 54.38$

$\hat{\gamma}_{11}^{AC} = \dfrac{315 - 612 + 584 - 967 - 453 + 710 - 737 + 1107}{24} = 2.21;$

$\hat{\gamma}_{21}^{AC} = -\hat{\gamma}_{11}^{AC} = 2.21.$

b. Factor SS's appear above. With $CF = \dfrac{5485^2}{24} = 1,253,551.04$ and

$\Sigma\Sigma\Sigma\Sigma x_{ijkl}^2 = 1,411,889$, SST = 158,337.96, from which SSE = 2608.7. The ANOVA table appears in the answer section. $F_{.05,1,16} = 4.49$, from which we see that the AB interaction and al the main effects are significant.

41. $\Sigma\Sigma\Sigma\Sigma\Sigma x^2_{ijklm} = 3{,}308{,}143$, $x_{\ldots\ldots} = 11{,}956$, so $CF = \dfrac{(11{,}956)^2}{48} = 2{,}979{,}535.02$, and SST

$= 328{,}607.98$. Each SS is $\dfrac{(\text{effect contrast})^2}{48}$ and SSE is obtained by subtraction. The

ANOVA table appears in the answer section (see back of textbook). $F_{.05,1,32} \approx 4.15$, a

value exceeded by the F ratios for AB interaction and the four main effects.

43.

Condition/ Effect	$SS = \frac{(contrast)^2}{16}$	f	Condition/ Effect	$SS = \frac{(contrast)^2}{16}$	f
(1)	—		D	414.123	1067.33
A	.436	1.12	AD	.017	< 1
B	.099	< 1	BD	.456	< 1
AB	.497	1.28	ABD	.055	—
C	.109	< 1	CD	2.190	5.64
AC	.078	< 1	ACD	1.020	—
BC	1.404	3.62	BCD	.133	—
ABC	.051	—	ABCD	.681	—

SSE = .051 + .055 + 1.020 + .133 + .681 = 1.940, df = 5, so MSE = .388.
$F_{.05,1,5} = 6.61$, so only the D main effect is significant.

45.

a. The allocation of treatments to blocks is as given in the answer section, with block #1 containing all treatments having an even number of letters in common with both ab and cd, etc.

b. $x_{\ldots\ldots} = 16{,}898$, so $SST = 9{,}035{,}054 - \dfrac{16{,}898^2}{32} = 111{,}853.88$. The eight block-replication totals are 2091 (= 618 + 421 + 603 + 449, the sum of the four observations in block #1 on replication #1), 2092, 2133, 2145, 2113, 2080, 2122, and 2122, so $SSBl = \dfrac{2091^2}{4} + \ldots + \dfrac{2122^2}{4} - \dfrac{16{,}898^2}{32} = 898.88$. The remaining SS's as well as all F ratios appear in the ANOVA table in the answer section. With $F_{.01,1,12} = 9.33$, only the A and B main effects are significant.

47. See the text's answer section.

49.

		A	B	C	D	E	AB	AC	AD	AE	BC	BD	BE	CD	CE	DE
a	70.4	+	−	−	−	−	−	−	−	−	+	+	+	+	+	+
b	72.1	−	+	−	−	−	−	+	+	+	−	−	−	+	+	+
c	70.4	−	−	+	−	−	+	−	+	+	−	+	+	−	−	+
abc	73.8	+	+	+	−	−	+	+	−	−	+	−	−	−	−	+
d	67.4	−	−	−	+	−	+	+	−	+	+	−	+	−	+	−
abd	67.0	+	+	−	+	−	+	−	+	−	−	+	−	−	+	−
acd	66.6	+	−	+	+	−	−	+	+	−	−	−	+	+	−	−
bcd	66.8	−	+	+	+	−	−	−	−	+	+	+	−	+	−	−
e	68.0	−	−	−	−	+	+	+	+	−	+	+	−	+	−	−
abe	67.8	+	+	−	−	+	+	−	−	+	−	−	+	+	−	−
ace	67.5	+	−	+	−	+	−	+	−	+	−	+	−	−	+	−
bce	70.3	−	+	+	−	+	−	−	+	−	+	−	+	−	+	−
ade	64.0	+	−	−	+	+	−	−	+	+	+	−	−	−	−	+
bde	67.9	−	+	−	+	+	−	+	−	−	−	+	+	−	−	+
cde	65.9	−	−	+	+	+	+	−	−	−	−	−	−	+	+	+
abcde	68.0	+	+	+	+	+	+	+	+	+	+	+	+	+	+	+

Thus $SSA = \dfrac{(70.4 - 72.1 - 70.4 + \ldots + 68.0)^2}{16} = 2.250$, SSB = 7.840, SSC = .360, SSD = 52.563, SSE = 10.240, SSAB = 1.563, SSAC = 7.563, SSAD = .090, SSAE = 4.203, SSBC = 2.103, SSBD = .010, SSBE = .123, SSCD = .010, SSCE = .063, SSDE = 4.840, Error SS = sum of two factor SS's = 20.568, Error MS = 2.057, $F_{.01,1,10} = 10.04$, so only the D main effect is significant.

Supplementary Exercises

51.

Source	df	SS	MS	f
A	1	322.667	322.667	980.38
B	3	35.623	11.874	36.08
AB	3	8.557	2.852	8.67
Error	16	5.266	.329	
Total	23	372.113		

We first test the null hypothesis of no interactions ($H_0 : \gamma_{ij} = 0$ for all i, j). H_0 will

be rejected at level .05 if $f_{AB} = \dfrac{MSAB}{MSE} \geq F_{.05,3,16} = 3.24$. Because $8.67 \geq 3.24$, H_0 is

rejected. Because we have concluded that interaction is present, tests for main effects are not appropriate.

53. Let A = spray volume, B = belt speed, C = brand.

Condition	Total	1	2	Contrast	$SS = \frac{(contrast)^2}{16}$
(1)	76	129	289	592	21,904.00
A	53	160	303	22	30.25
B	62	143	13	48	144.00
AB	98	160	9	134	1122.25
C	88	−23	31	14	12.25
AC	55	36	17	−4	1.00
BC	59	−33	59	−14	12.25
ABC	101	42	75	16	16.00

The ANOVA table is as follows:

Effect	Df	MS	f
A	1	30.25	6.72
B	1	144.00	32.00
AB	1	1122.25	249.39
C	1	12.25	2.72
AC	1	1.00	.22
BC	1	12.25	2.72
ABC	1	16.00	3.56
Error	8	4.50	
Total	15		

$F_{.05,1,8} = 5.32$, so all of the main effects are significant at level .05, but none of the interactions are significant.

55.

a.

Effect	%Iron	1	2	3	Effect Contrast	SS
	7	18	37	174	684	
A	11	19	137	510	144	1296
B	7	62	169	50	36	81
AB	12	75	341	94	0	0
C	21	79	9	14	272	4624
AC	41	90	41	22	32	64
BC	27	165	47	2	12	9
ABC	48	176	47	−2	−4	1
D	28	4	1	100	336	7056
AD	51	5	13	172	44	121
BD	33	20	11	32	8	4
ABD	57	21	11	0	0	0
CD	70	23	1	12	72	324
ACD	95	24	1	0	−32	64
BCD	77	25	1	0	−12	9
ABCD	99	22	−3	−4	−4	1

We use $estimate = \dfrac{contrast}{2^p}$ when $n = 1$ to get $\hat{\alpha}_1 = \dfrac{144}{2^4} = \dfrac{144}{16} = 9.00$,

$\hat{\beta}_1 = \dfrac{36}{16} = 2.25$, $\hat{\delta}_1 = \dfrac{272}{16} = 17.00$, $\hat{\gamma}_1 = \dfrac{336}{16} = 21.00$. Similarly, $\left(\widehat{\alpha\beta}\right)_{11} = 0$,

$\left(\widehat{\alpha\delta}\right)_{11} = 2.00$, $\left(\widehat{\alpha\gamma}\right)_{11} = 2.75$, $\left(\widehat{\beta\delta}\right)_{11} = .75$, $\left(\widehat{\beta\gamma}\right)_{11} = .50$, and $\left(\widehat{\delta\gamma}\right)_{11} = 4.50$.

b.

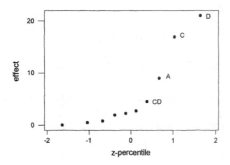

The plot suggests main effects A, C, and D are quite important, and perhaps the interaction CD as well. (See answer section for comment.)

57. The ANOVA table is:

Source	df	SS	MS	f	$F_{.01}$
A	2	34,436	17,218	436.92	5.49
B	2	105,793	52,897	1342.30	5.49
C	2	516,398	258,199	6552.04	5.49
AB	4	6,868	1,717	43.57	4.11
AC	4	10,922	2,731	69.29	4.11
BC	4	10,178	2,545	64.57	4.11
ABC	8	6,713	839	21.30	3.26
Error	27	1,064	39		
Total	53	692,372			

All calculated F values are greater than their respective tabled values, so all effects, including the interaction effects, are significant at level .01.

59. Based on the P-values in the ANOVA table, statistically significant factors at the level .01 are adhesive type and cure time. The conductor material does not have a statistically significant effect on bond strength. There are no significant interactions.

61. $SSA = \sum_i \sum_j \left(\overline{X}_{i...} - \overline{X}_{....} \right)^2 = \frac{1}{N} \Sigma X_{i...}^2 - \frac{X_{....}^2}{N}$, with similar expressions for SSB, SSC, and SSD, each having N − 1 df.

$SST = \sum_i \sum_j \left(X_{ij(kl)} - \overline{X}_{....} \right)^2 = \sum_i \sum_j X_{ij(kl)}^2 - \frac{X_{....}^2}{N}$ with $N^2 - 1$ df, leaving

$N^2 - 1 - 4(N - 1)$ df for error.

	1	2	3	4	5	Σx^2
$x_{i...}$:	482	446	464	468	434	1,053,916
$x_{.j..}$:	470	451	440	482	451	1,053,626
$x_{..k.}$:	372	429	484	528	481	1,066,826
$x_{...l}$:	340	417	466	537	534	1,080,170

Also, $\Sigma\Sigma x_{ij(kl)}^2 = 220,378$, $x_{....} = 2294$, and CF = 210,497.44

Source	df	SS	MS	f	$F_{.05}$
A	4	285.76	71.44	.594	3.84
B	4	227.76	56.94	.473	3.84
C	4	2867.76	716.94	5.958*	3.84
D	4	5536.56	1384.14	11.502*	3.84
Error	8	962.72	120.34		
Total	24				

H_{0A} and H_{0B} cannot be rejected, while while H_{0C} and H_{0D} are rejected.

CHAPTER 12

Section 12.1

1.

    a.   Stem and Leaf display of temp:

```
17|0
17|23           stem = tens
17|445          leaf = ones
17|67
17|
18|0000011
18|2222
18|445
18|6
18|8
```

180 appears to be a typical value for this data. The distribution is reasonably symmetric in appearance and somewhat bell-shaped. The variation in the data is fairly small since the range of values (188 − 170 = 18) is fairly small compared to the typical value of 180.

```
0|889
1|0000          stem = ones
1|3             leaf = tenths
1|4444
1|66
1|8889
2|11
2|
2|5
2|6
2|
3|00
```

For the ratio data, a typical value is around 1.6 and the distribution appears to be positively skewed. The variation in the data is large since the range of the data (3.08 − .84 = 2.24) is very large compared to the typical value of 1.6. The two largest values could be outliers.

b. The efficiency ratio is not uniquely determined by temperature since there are several instances in the data of equal temperatures associated with different efficiency ratios. For example, the five observations with temperatures of 180 each have different efficiency ratios.

c. A scatter plot of the data appears below. The points exhibit quite a bit of variation and do not appear to fall close to any straight line or simple curve.

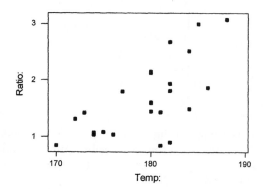

3. A scatter plot of the data appears below. The points fall very close to a straight line with an intercept of approximately 0 and a slope of about 1. This suggests that the two methods are producing substantially the same concentration measurements.

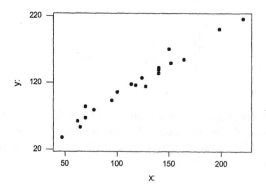

5.

a. The scatter plot with axes intersecting at (0,0) is shown below.

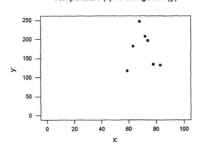

Temperature (x) vs Elongation (y)

b. The scatter plot with axes intersecting at (55, 100) is shown below.

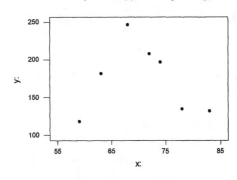

Temperature (x) vs Elongation (y)

c. A parabola appears to provide a good fit to both graphs.

7.

a. $\mu_{Y \cdot 2500} = 1800 + 1.3(2500) = 5050$

b. expected change = slope = $\beta_1 = 1.3$

c. $100\beta_1 = 130$

d. $-100\beta_1 = -130$

Chapter 12: Simple Linear Regression and Correlation

9.

    a. β_1 = expected change in flow rate (y) associated with a one inch increase in pressure drop (x) = .095.

    b. We expect flow rate to decrease by $5\beta_1 = .475$.

    c. $\mu_{Y\cdot10} = -.12 + .095(10) = .83$, and $\mu_{Y\cdot15} = -.12 + .095(15) = 1.305$.

    d. $P(Y > .835) = P\left(Z > \dfrac{.835 - .830}{.025}\right) = P(Z > .20) = .4207$;

        $P(Y > .840) = P\left(Z > \dfrac{.840 - .830}{.025}\right) = P(Z > .40) = .3446$

    e. Let Y_1 and Y_2 denote pressure drops for flow rates of 10 and 11, respectively. Then $\mu_{Y\cdot11} = .925$, so $Y_1 - Y_2$ has expected value $.830 - .925 = -.095$, and

        $SD(Y_1 - Y_2) = \sqrt{(.025)^2 + (.025)^2} = .035355$. Thus,

        $P(Y_1 > Y_2) = P(Y_1 - Y_2 > 0) = P\left(z > \dfrac{+.095}{.035355}\right) = P(Z > 2.69) = .0036$

11.

    a. β_1 = expected change for a one degree increase = $-.01$, and $10\beta_1 = -.1$ is the expected change for a 10 degree increase.

    b. $\mu_{Y\cdot200} = 5.00 - .01(200) = 3$, and $\mu_{Y\cdot250} = 2.5$.

    c. The probability that the first observation is between 2.4 and 2.6 is

        $P(2.4 \le Y \le 2.6) = P\left(\dfrac{2.4 - 2.5}{.075} \le Z \le \dfrac{2.6 - 2.5}{.075}\right) = P(-1.33 \le Z \le 1.33) = .8164$.

        The probability that any particular one of the other four observations is between 2.4 and 2.6 is also .8164, so the probability that all five are between 2.4 and 2.6 is $(.8164)^5 = .3627$.

    d. Let Y_1 and Y_2 denote the times at the higher and lower temperatures, respectively. Then $Y_1 - Y_2$ has expected value
        $5.00 - .01(x + 1) - (5.00 - .01x) = -.01$. The standard deviation of $Y_1 - Y_2$ is

        $\sqrt{(.075)^2 + (.075)^2} = .10607$. Thus,

        $P(Y_1 - Y_2 > 0) = P\left(z > \dfrac{-(-.01)}{.10607}\right) = P(Z > .09) = .4641$.

Section 12.2

13. For this data, $n = 4$, $\Sigma x_i = 200$, $\Sigma y_i = 5.37$, $\Sigma x_i^2 = 12.000$, $\Sigma y_i^2 = 9.3501$,

$\Sigma x_i y_i = 333$. From these, $S_{xx} = 12,000 - \dfrac{(200)^2}{4} = 2000$,

$S_{yy} = 9.3501 - \dfrac{(5.37)^2}{4} = 2.140875$, and $S_{xy} = 333 - \dfrac{(200)(5.37)}{4} = 64.5$.

$\hat{\beta}_1 = \dfrac{S_{xy}}{S_{xx}} = \dfrac{64.5}{2000} = .03225$ and $\hat{\beta}_0 = \dfrac{5.37}{4} - (.03225)\dfrac{200}{4} = -.27000$.

$SSE = S_{yy} - \hat{\beta}_1 S_{xy} = 2.14085 - (.03225)(64.5) = .060750$.

$r^2 = 1 - \dfrac{SSE}{SST} = 1 - \dfrac{.060750}{2.14085} = .972$. This is a very high value of r^2, which confirms
the authors' claim that there is a strong linear relationship between the two variables.

15.

a. The following stem and leaf display shows that: a typical value for this data is a
number in the low 40's. There is some positive skew in the data. There are some
potential outliers (79.5 and 80.0), and there is a reasonably large amount of
variation in the data (e.g., the spread 80.0–29.8 = 50.2 is large compared with the
typical values in the low 40's).

```
2|9
3|33                    stem = tens
3|5566677889            leaf = ones
4|1223
4|56689
5|1
5|
6|2
6|9
7|
7|9
8|0
```

b. No, the strength values are not uniquely determined by the MoE values. For
example, note that the two pairs of observations having strength values of 42.8
have different MoE values.

c. The least squares line is $\hat{y} = 3.2925 + .10748x$. For a beam whose modulus of elasticity is x = 40, the predicted strength would be $\hat{y} = 3.2925 + .10748(40) = 7.59$. The value x = 100 is far beyond the range of the x values in the data, so it would be dangerous (i.e., potentially misleading) to extrapolated the linear relationship that far.

d. From the output, SSE = 18.736, SST = 71.605, and the coefficient of determination is $r^2 = .738$ (or 73.8%). The r^2 value is large, which suggests that the linear relationship is a useful approximation to the true relationship between these two variables.

17. Note: $n = 23$ in this study.

a. For a one (mg/cm²) increase in dissolved material, one would expect a .144 (g/l) increase in calcium content. Secondly, 86% of the observed variation in calcium content can be attributed to the simple linear regression relationship between calcium content and dissolved material.

b. $\mu_{y \cdot 50} = 3.678 + .144(50) = 10.878$

c. $r^2 = .86 = 1 - \dfrac{SSE}{SST}$, so $SSE = (SST)(1 - .86) = (320.398)(.14) = 44.85572$. Then

$$s = \sqrt{\dfrac{SSE}{n-2}} = \sqrt{\dfrac{44.85572}{21}} = 1.46$$

19. $n = 14$, $\Sigma x_i = 3300$, $\Sigma y_i = 5010$, $\Sigma x_i^2 = 913,750$, $\Sigma y_i^2 = 2,207,100$, $\Sigma x_i y_i = 1,413,500$

a. $\hat{\beta}_1 = \dfrac{3,256,000}{1,902,500} = 1.71143233$, $\hat{\beta}_0 = -45.55190543$, so we use the equation $\hat{y} = -45.5519 + 1.7114x$.

b. $\hat{\mu}_{Y \cdot 225} = -45.5519 + 1.7114(225) = 339.51$

c. Estimated expected change $= -50\hat{\beta}_1 = -85.57$

d. No, the value 500 is outside the range of x values for which observations were available (the danger of extrapolation).

Chapter 12: Simple Linear Regression and Correlation

21.

    a.   Yes – a scatter plot of the data shows a strong, linear pattern, and $r^2 = 98.5\%$.

    b.   From the output, the estimated regression line is $\hat{y} = 321.878 + 156.711x$, where x = absorbance and y = resistance angle. For $x = .300$, $\hat{y} = 321.878 + 156.711(.300) = 368.89$.

    c.   The estimated regression line serves as an estimate both for a single y at a given x-value and for the true average μ_y at a given x-value. Hence, our estimate for μ_y when $x = .300$ is also 368.89.

23.

    a.   Using the given y_i's and the formula $\hat{y}_i = -45.5519 + 1.7114x_i$,

$$SSE = (150 - 125.6)^2 + \ldots + (670 - 639.0)^2 = 16{,}213.64.$$ The computation formula gives $SSE = 2{,}207{,}100 - (-45.55190543)(5010) - (1.71143233)(1{,}413{,}500)$
$$= 16{,}205.45$$

    b. $SST = 2{,}207{,}100 - \dfrac{(5010)^2}{14} = 414{,}235.71$ so $r^2 = 1 - \dfrac{16{,}205.45}{414{,}235.71} = .961$.

25.   Substitution of $\hat{\beta}_0 = \dfrac{\Sigma y_i - \hat{\beta}_1 \Sigma x_i}{n}$ and $\hat{\beta}_1$ for b_0 and b_1 on the left hand side of the normal equations yields $\dfrac{n(\Sigma y_i - \hat{\beta}_1 \Sigma x_i)}{n} + \hat{\beta}_1 \Sigma x_i = \Sigma y_i$ from the first equation and

$$\frac{\Sigma x_i (\Sigma y_i - \hat{\beta}_1 \Sigma x_i)}{n} + \hat{\beta}_1 \Sigma x_i^2 = \frac{\Sigma x_i \Sigma y_i}{n} + \frac{\hat{\beta}_1 (n \Sigma x_i^2 - (\Sigma x_i)^2)}{n}$$

$$\frac{\Sigma x_i \Sigma y_i}{n} + \frac{n \Sigma x_i y_i}{n} - \frac{\Sigma x_i \Sigma y_i}{n} = \Sigma x_i y_i$$ from the second equation.

27.   We wish to find b_1 to minimize $f(b_1) = \Sigma(y_i - b_1 x_i)^2$. Equating $f'(b_1)$ to 0 yields $2\Sigma(y_i - b_1 x_i)(-x_i) = 0$ so $\Sigma x_i y_i = b_1 \Sigma x_i^2$ and $b_1 = \dfrac{\Sigma x_i y_i}{\Sigma x_i^2}$. The least squares estimator of $\hat{\beta}_1$ is thus $\hat{\beta}_1 = \dfrac{\Sigma x_i Y_i}{\Sigma x_i^2}$.

29. For data set #1, $r^2 = .43$ and $\hat{\sigma} = s = 4.03$; whereas these quantities are .99 and 4.03 for #2, and .99 and 1.90 for #3. In general, one hopes for both large r^2 (large % of variation explained) and small s (indicating that observations don't deviate much from the estimated line). Simple linear regression would thus seem to be most effective in the third situation.

Section 12.3

31.

a. $\hat{\beta}_1 = -.00736023$ and $\hat{\beta}_0 = 1.41122185$, so

$SSE = 7.8518 - (1.41122185)(10.68) - (-.00736023)(987.645) = .04925$, from

which $s^2 = .003788$ and $s = .06155$.

$$\hat{\sigma}_{\hat{\beta}_1}^2 = \frac{s^2}{\Sigma x_i^2 - (\Sigma x_i)^2 / n} = \frac{.003788}{3662.25} = .00000103 \text{, from which}$$

$\hat{\sigma}_{\hat{\beta}_1} = s_{\hat{\beta}_1} = $ estimated sd of $\hat{\beta}_1 = \sqrt{.00000103} = .001017$.

b. $-.00736 \pm (2.160)(.001017) = -.00736 \pm .00220 = (-.00956, -.00516)$

33.

a. Error d.f. $= n - 2 = 25$, $t_{.025,25} = 2.060$, and so the desired confidence interval is

$\hat{\beta}_1 \pm t_{.025,25} \cdot s_{\hat{\beta}_1} = .10748 \pm (2.060)(.01280) = (.081, .134)$. We are 95% confident

that the true average change in strength associated with a 1 GPa increase in modulus of elasticity is between .081 MPa and .134 MPa.

b. We wish to test $H_0 : \beta_1 \leq .1$ vs. $H_a : \beta_1 > .1$. The calculated t statistic is

$t = \dfrac{\hat{\beta}_1 - .1}{s_{\hat{\beta}_1}} = \dfrac{.10748 - .1}{.01280} = .58$, which yields a P-value of .277 at 25 d.f. Thus, we

fail to reject H_0; i.e., there is not enough evidence to contradict the prior belief.

35.

a. We want a 95% CI for β_1: $\hat{\beta}_1 \pm t_{.025,15} s_{\hat{\beta}_1}$. Using the given summary statistics,

$$S_{xx} = 3056.69 - \frac{(222.1)^2}{17} = 155.019, \quad S_{xy} = 2759.6 - \frac{(222.1)(193)}{17} = 238.112,$$

and $\hat{\beta}_1 = \dfrac{S_{xy}}{S_{xx}} = \dfrac{238.112}{115.019} = 1.536$. We need $\hat{\beta}_0 = \dfrac{193 - (1.536)(222.1)}{17} = -8.715$

to calculate the SSE: $SSE = 2975 - (-8.715)(193) - (1.536)(2759.6) = 418.2494$.

Then $s = \sqrt{\dfrac{418.2494}{15}} = 5.28$ and $s_{\hat{\beta}_1} = \dfrac{5.28}{\sqrt{155.019}} = .424$. With $t_{.025,15} = 2.131$,

our CI is $1.536 \pm 2.131(.424) = (.632, 2.440)$. With 95% confidence, we estimate that the change in reported nausea percentage for every one-unit change in motion sickness dose is between .632 and 2.440.

b. We test the hypotheses H_0: $\beta_1 = 0$ v. H_a: $\beta_1 \neq 0$, and the test statistic is $t = \dfrac{1.536}{.424} = 3.6226$. With df = 15, the two-tailed P-value = $2P(t > 3.6226) = 2(.001) = .002$. With a P-value of .002, we would reject the null hypothesis at most reasonable significance levels. This suggests that there is a useful linear relationship between motion sickness dose and reported nausea.

c. No. A regression model is only useful for estimating values of nausea % when using dosages between 6.0 and 17.6 – the range of values sampled.

d. Removing the point (6.0, 2.50), the new summary stats are: $n = 16$, $\Sigma x_i = 216.1$, $\Sigma y_i = 191.5$, $\Sigma x_i^2 = 3020.69$, $\Sigma y_i^2 = 2968.75$, $\Sigma x_i y_i = 2744.6$, and then $\hat{\beta}_1 = 1.561$, $\hat{\beta}_0 = -9.118$, $SSE = 430.5264$, $s = 5.55$, $s_{\hat{\beta}_1} = .551$, and the new CI is $1.561 \pm 2.145 \cdot (.551)$, or $(.379, 2.743)$. The interval is a little wider. But removing the one observation did not change it that much. The observation does not seem to be exerting undue influence.

37.

a. Let μ_d = the true mean difference in velocity between the two planes. We have 23 pairs of data that we will use to test H_0: $\mu_d = 0$ v. H_a: $\mu_d \neq 0$. From software, $\bar{x}_d = 0.2913$ with $s_d = 0.1748$, and so $t = \dfrac{0.2913 - 0}{0.1748} \approx 8$, which has a two-sided P-value of 0.000 at 22 df. Hence, we strongly reject the null hypothesis and conclude there is a statistically significant difference in true average velocity in the two planes. [*Note:* A normal probability plot of the differences shows one mild outlier, so we have slight concern about the results of the t procedure.]

b. Let β_1 denote the true slope for the linear relationship between Level – – velocity and Level – velocity. We wish to test H_0: $\beta_1 = 1$ v. H_a: $\beta_1 < 1$. Using the relevant numbers provided, $t = \dfrac{b_1 - 1}{s(b_1)} = \dfrac{0.65393 - 1}{0.05947} = -5.8$, which has a one-sided P-value at 23–2 = 21 df of $P(t < -5.8) \approx 0$. Hence, we strongly reject the null hypothesis and conclude the same as the authors; i.e., the true slope of this regression relationship is significantly less than 1.

39. SSE = 124,039.58– (72.958547)(1574.8) – (.04103377)(222657.88) = 7.9679, and SST = 39.828

Source	df	SS	MS	f
Regr	1	31.860	31.860	18.0
Error	18	7.968	1.77	
Total	19	39.828		

Let's use $\alpha = .001$. Then $F_{.001,1,18} = 15.38 < 18.0$, so H_0: $\beta_1 = 0$ is rejected and the model is judged useful. $s = \sqrt{1.77} = 1.33041347$ and $S_{xx} = 18,921.8295$, so $t = \dfrac{.04103377}{1.33041347 / \sqrt{18,921.8295}} = 4.2426$, and $t^2 = (4.2426)^2 = 18.0 = f$.

41.

a. Under the regression model, $E(Y_i) = \beta_0 + \beta_1 x_i$ and, hence, $E(\bar{Y}) = \beta_0 + \beta_1 \bar{x}$. Therefore, $E(Y_i - \bar{Y}) = \beta_1(x_i - \bar{x})$, and

$$E(\hat{\beta}_1) = E\left[\frac{\sum(x_i - \bar{x})(Y_i - \bar{Y})}{\sum(x_i - \bar{x})^2}\right] = \frac{\sum(x_i - \bar{x})E[Y_i - \bar{Y}]}{\sum(x_i - \bar{x})^2}$$

$$= \frac{\sum(x_i - \bar{x})\beta_1(x_i - \bar{x})}{\sum(x_i - \bar{x})^2} = \beta_1 \frac{\sum(x_i - \bar{x})^2}{\sum(x_i - \bar{x})^2} = \beta_1.$$

b. With $c = \Sigma(x_i - \bar{x})^2$, $\hat{\beta}_1 = \frac{1}{c}\Sigma(x_i - \bar{x})(Y_i - \bar{Y}) = \frac{1}{c}\Sigma(x_i - \bar{x})Y_i$ (since

$\Sigma(x_i - \bar{x})\bar{Y} = \bar{Y}\Sigma(x_i - \bar{x}) = \bar{Y}\cdot 0 = 0$), so $V(\hat{\beta}_1) = \frac{1}{c^2}\Sigma(x_i - \bar{x})^2 Var(Y_i)$

$$= \frac{1}{c^2}\Sigma(x_i - \bar{x})^2 \cdot \sigma^2 = \frac{\sigma^2}{\Sigma(x_i - \bar{x})^2} = \frac{\sigma^2}{\Sigma x_i^2 - (\Sigma x_i)^2/n}, \text{ as desired.}$$

43.   The numerator of d is $|1 - 2| = 1$, and the denominator is $\dfrac{4\sqrt{14}}{\sqrt{324.40}} = .831$, so

$d = \dfrac{1}{.831} = 1.20$. The approximate power curve is for $n-2$ df $= 13$, and β is read from Table A.17 as approximately .1.

Section 12.4

45.

    a. We wish to find a 90% CI for $\mu_{y \cdot 125}$: $\hat{y}_{125} = 78.088$, $t_{.05,18} = 1.734$, and

$$s_{\hat{y}} = s\sqrt{\frac{1}{20} + \frac{(125 - 140.895)^2}{18,921.8295}} = .1674.$$ Putting it together, we get

$$78.088 \pm 1.734(.1674) = (77.797, 78.378).$$

    b. We want a 90% PI: Only the standard error changes:

$$s_{\hat{y}} = s\sqrt{1 + \frac{1}{20} + \frac{(125 - 140.895)^2}{18,921.8295}} = .6860,$$ so the PI is

$$78.088 \pm 1.734(.6860) = (76.898, 79.277).$$

    c. Because the x* of 115 is farther away from $\bar{x}$ than the previous value, the term $(x^* - \bar{x})^2$ will be larger, making the standard error larger, and thus the width of the interval is wider.

    d. We would be testing to see: if the filtration rate were 125 kg-DS/m/h, would the average moisture content of the compressed pellets be less than 80%. The test statistic is $t = \dfrac{78.088 - 80}{.1674} = -11.42$, and with 18 df the P-value is $P(t < -11.42) \approx 0.00$. We would reject H_0. There is significant evidence to prove that the true average moisture content when filtration rate is 125 is less than 80%.

47.

    a. $\hat{y}_{(40)} = -1.128 + .82697(40) = 31.95$, $t_{.025,13} = 2.160$; a 95% PI for runoff is

$$31.95 \pm 2.160\sqrt{(5.24)^2 + (1.44)^2} = 31.95 \pm 11.74 = (20.21, 43.69).$$ No, the resulting interval is very wide; therefore, the available information is not very precise.

    b. $\Sigma x = 798$, $\Sigma x^2 = 63,040$ give $S_{xx} = 20,586.4$, which in turn gives

$$s_{\hat{y}(50)} = 5.24\sqrt{\frac{1}{15} + \frac{(50 - 53.20)^2}{20,586.4}} = 1.358,$$ so the PI for runoff when x = 50 is

$$40.22 \pm 2.160\sqrt{(5.24)^2 + (1.358)^2} = 40.22 \pm 11.69 = (28.53, 51.92).$$ The simultaneous prediction level for the two intervals is at least $100(1 - 2\alpha)\% = 90\%$.

49. 95% CI: $(462.1, 597.7)$; midpoint $= 529.9$; $t_{.025,8} = 2.306$;

$$529.9 + (2.306)\left(\hat{s}_{\hat{\beta}_0 + \hat{\beta}_1(15)}\right) = 597.7 \rightarrow \hat{s}_{\hat{\beta}_0 + \hat{\beta}_1(15)} = 29.402 \rightarrow 99\% \text{ CI}$$

$$= 529.9 \pm (3.355)(29.402) = (431.3, 628.5)$$

51.

   a.   0.40 is closer to $\bar{x}$.

   b. $\hat{\beta}_0 + \hat{\beta}_1(0.40) \pm t_{\alpha/2, n-2} \cdot \left(\hat{s}_{\hat{\beta}_0 + \hat{\beta}_1(0.40)}\right)$ or $0.8104 \pm (2.101)(0.0311)$

       $= (0.745, 0.876)$

   c. $\hat{\beta}_0 + \hat{\beta}_1(1.20) \pm t_{\alpha/2, n-2} \cdot \sqrt{s^2 + s^2_{\hat{\beta}_0 + \hat{\beta}_1(1.20)}}$ or

       $0.2912 \pm (2.101) \cdot \sqrt{(0.1049)^2 + (0.0352)^2} = (.059, .523)$

53. Choice **a** will be the smallest, with d being largest. **a** is less than **b** and **c** (obviously), and **b** and **c** are both smaller than **d**. Nothing can be said about the relationship between **b** and **c**.

55. $\hat{\beta}_0 + \hat{\beta}_1 x = \bar{Y} - \hat{\beta}_1 \bar{x} + \hat{\beta}_1 x = \bar{Y} + (x - \bar{x})\hat{\beta}_1 = \dfrac{1}{n}\sum Y_i + \dfrac{(x - \bar{x})\sum (x_i - \bar{x})Y_i}{S_{XX}} = \sum d_i Y_i$

where $d_i = \dfrac{1}{n} + \dfrac{(x - \bar{x})(x_i - \bar{x})}{S_{XX}}$. Thus, $Var\left(\hat{\beta}_0 + \hat{\beta}_1 x\right) = \sum d_i^2 Var(Y_i) = \sigma^2 \sum d_i^2$

$$= \sigma^2 \sum \left[\frac{1}{n^2} + 2\frac{(x - \bar{x})(x_i - \bar{x})}{nS_{xx}} + \frac{(x - \bar{x})^2 (x_i - \bar{x})^2}{nS_{xx}^2} \right]$$

$$= \sigma^2 \left[n\frac{1}{n^2} + 2\frac{(x - \bar{x})\sum (x_i - \bar{x})}{nS_{xx}} + \frac{(x - \bar{x})^2 \sum (x_i - \bar{x})^2}{S_{xx}^2} \right]$$

$$= \sigma^2 \left[\frac{1}{n} + 2\frac{(x - \bar{x}) \cdot 0}{nS_{xx}} + \frac{(x - \bar{x})^2 S_{XX}}{S_{xx}^2} \right] = \sigma^2 \left[\frac{1}{n} + \frac{(x - \bar{x})^2}{S_{XX}} \right]$$

Section 12.5

57. Most people acquire a license as soon as they become eligible. If, for example, the minimum age for obtaining a license is 16, then the time since acquiring a license, y, is usually related to age by the equation $y \approx x - 16$, which is the equation of a straight line. In other words, the majority of people in a sample will have y values that closely follow the line $y = x - 16$.

59.

a. $S_{xx} = 251,970 - \dfrac{(1950)^2}{18} = 40,720$, $S_{yy} = 130.6074 - \dfrac{(47.92)^2}{18} = 3.033711$, and

$S_{xy} = 5530.92 - \dfrac{(1950)(47.92)}{18} = 339.586667$, so

$r = \dfrac{339.586667}{\sqrt{40,720}\sqrt{3.033711}} = .9662$. There is a very strong positive correlation between the two variables.

b. Because the association between the variables is positive, the specimen with the larger shear force will tend to have a larger percent dry fiber weight.

c. Changing the units of measurement on either (or both) variables will have no effect on the calculated value of r, because any change in units will affect both the numerator and denominator of r by exactly the same multiplicative constant.

d. $r^2 = (.966)^2 = .933$

e. $H_0 : \rho = 0$ vs $H_a : \rho > 0$. $t = \dfrac{r\sqrt{n-2}}{\sqrt{1-r^2}}$; Reject H_0 at level .01 if

$t \geq t_{.01,16} = 2.583$. $t = \dfrac{.966\sqrt{16}}{\sqrt{1-.966^2}} = 14.94 \geq 2.583$, so H_0 should be rejected.

The data indicates a positive linear relationship between the two variables.

61.

a. We are testing $H_0 : \rho = 0$ vs $H_a : \rho > 0$.

$$r = \frac{7377.704}{\sqrt{36.9839}\sqrt{2,628,930.359}} = .7482, \text{ and } t = \frac{.7482\sqrt{12}}{\sqrt{1-.7482^2}} = 3.9066. \text{ We}$$

reject H_0 since $t = 3.9066 \geq t_{.05,12} = 1.782$. There is evidence that a positive correlation exists between maximum lactate level and muscular endurance.

b. We are looking for r^2, the coefficient of determination. $r^2 = (.7482)^2 = .5598$. It is the same no matter which variable is the predictor.

63. $n = 6$, $\Sigma x_i = 111.71, \Sigma x_i^2 = 2,724.7643, \Sigma y_i = 2.9, \Sigma y_i^2 = 1.6572$, and $\Sigma x_i y_i = 63.915$.

$$r = \frac{(6)(63.915) - (111.71)(2.9)}{\sqrt{(6)(2,724.7943) - (111.73)^2} \cdot \sqrt{(6)(1.6572) - (2.9)^2}} = .7729. \ H_0 : \rho = 0 \text{ v.}$$

$H_a : \rho \neq 0$; Reject H_0 at level .05 if $|t| \geq t_{.025,4} = 2.776$. $t = \frac{(.7729)\sqrt{4}}{\sqrt{1-(.7729)^2}} = 2.436$.

Fail to reject H_0. The data does not indicate that the population correlation coefficient differs from 0. This result may seem surprising due to the relatively large size of r (.77), however, it can be attributed to a small sample size (6).

65.

a. Although the normal probability plot of the x's appears somewhat curved, such a pattern is not terribly unusual when n is small; the test of normality presented in section 14.2 does not reject the hypothesis of population normality. The normal probability plot of the y's is much straighter.

b. $H_0 : \rho = 0$ will be rejected in favor of $H_a : \rho \neq 0$ at level .01 if

$|t| \geq t_{.005,8} = 3.355$. $\Sigma x_i = 864, \Sigma x_i^2 = 78,142, \Sigma y_i = 138.0, \Sigma y_i^2 = 1959.1$ and

$\Sigma x_i y_i = 12,322.4$, so $r = \frac{3992}{(186.8796)(23.3880)} = .913$ and

$t = \frac{.913(2.8284)}{.4080} = 6.33 \geq 3.355$, so reject H_0. There does appear to be a linear relationship.

67.

a. Because P-value = .00032 < α = .001, H_o should be rejected at this significance level.

b. Not necessarily. For this n, the test statistic t has approximately a standard normal distribution when $H_0 : \rho = 0$ is true, and a P-value of .00032 corresponds to $z = 3.60$ (or −3.60). Solving $3.60 = \dfrac{r\sqrt{498}}{\sqrt{1-r^2}}$ for r yields $r =$ .159. This r suggests only a weak linear relationship between x and y, one that would typically have little practical importance.

c. $t = \dfrac{.022\sqrt{9998}}{\sqrt{1-.022^2}} = 2.20 \geq t_{.025,9998} = 1.96$, so H_0 is rejected in favor of H_a. The value $t = 2.20$ is statistically significant – it cannot be attributed just to sampling variability in the case $\rho = 0$. But with this n, $r = .022$ implies $\rho \approx .022$, which in turn shows an extremely weak linear relationship.

Supplementary Exercises

69. Use available software for all calculations.

a. We want a confidence interval for β_1. From software, $b_1 = 0.987$ and $s(b_1) =$ 0.047, so the corresponding 95% CI is $0.987 \pm t_{.025,17}(0.047) = 0.987 \pm$ 2.110(0.047) = (0.888, 1.086). We are 95% confident that the true average change in sale price associated with a one-foot increase in truss height is between $0.89 per square foot and $1.09 per square foot.

b. Using software, a 95% CI for $\mu_{y.25}$ is (47.730, 49.172). We are 95% confident that the true average sale price for all warehouses with 25-foot truss height is between $47.73/ft^2 and $49.17/ft^2.

c. Again using software, a 95% PI for Y when x = 25 is (45.378, 51.524). We are 95% confident that the sale price for a single warehouse with 25-foot truss height will be between $45.38/ft^2 and $51.52/ft^2.

d. Since x = 25 is nearer the mean than x = 30, a PI at x = 30 would be wider.

e. From software, r^2 = SSR/SST = 890.36/924.44 = .963. Hence, $r = \sqrt{.963} = .981$.

71.

a. The test statistic value is $t = \dfrac{\hat{\beta}_1 - 1}{s_{\hat{\beta}_1}}$, and H_0 will be rejected if either

$t \geq t_{.025,11} = 2.201$ or $t \leq -2.201$.

With $\Sigma x_i = 243, \Sigma x_i^2 = 5965, \Sigma y_i = 241, \Sigma y_i^2 = 5731$ and $\Sigma x_i y_i = 5805$,

$\hat{\beta}_1 = .913819$, $\hat{\beta}_0 = 1.457072$, $SSE = 75.126$, $s = 2.613$, and $s_{\hat{\beta}_1} = .0693$,

$t = \dfrac{.9138 - 1}{.0693} = -1.24$. Because -1.24 is neither ≤ -2.201 nor ≥ 2.201, H_0

cannot be rejected. It is plausible that $\beta_1 = 1$.

b. $r = \dfrac{16,902}{(136)(128.15)} = .970$

73.

a. $r^2 = .5073$

b. $r = +\sqrt{r^2} = \sqrt{.5073} = .7122$ (positive because $\hat{\beta}_1$ is positive)

c. We test $H_0 : \beta_1 = 0$ v. $H_a : \beta_1 \neq 0$. The test statistic t = 3.93 gives P-value = .0013, which is < .01, the given level of significance, therefore we reject H_0 and conclude that the model is useful.

d. We use a 95% CI for $\mu_{Y \cdot 50}$. $\hat{y}_{(50)} = .787218 + .007570(50) = 1.165718$,

$t_{.025,15} = 2.131$, s = "Root MSE" = .020308, so

$s_{\hat{y}_{(50)}} = .20308\sqrt{\dfrac{1}{17} + \dfrac{17(50 - 42.33)^2}{17(41,575) - (719.60)^2}} = .051422$. The interval is, then,

$1.165718 \pm 2.131(.051422) = 1.165718 \pm .109581 = (1.056137, 1.275299)$.

e. $\hat{y}_{(30)} = .787218 + .007570(30) = 1.0143$. The residual is

$y - \hat{y} = .80 - 1.0143 = -.2143$.

75.

a. $n = 9$, $\Sigma x_i = 228$, $\Sigma x_i^2 = 5958$, $\Sigma y_i = 93.76$, $\Sigma y_i^2 = 982.2932$ and

$\Sigma x_i y_i = 2348.15$, giving $\hat{\beta}_1 = \dfrac{-243.93}{1638} = -.148919$, $\hat{\beta}_0 = 14.190392$, and the

equation $\hat{y} = 14.19 - .1489x$.

b. β_1 is the expected increase in load associated with a one-day age increase (so a negative value of β_1 corresponds to a decrease). We wish to test $H_0 : \beta_1 = -.10$ v. $H_a : \beta_1 < -.10$ (the alternative contradicts prior belief). H_0 will be rejected at

level .05 if $t = \dfrac{\hat{\beta}_1 - (-.10)}{s_{\hat{\beta}_1}} \leq -t_{.05,7} = -1.895$. With SSE = 1.4862, s = .4608,

and $s_{\hat{\beta}_1} = \dfrac{.4608}{\sqrt{182}} = .0342$. Thus $t = \dfrac{-.1489 + .1}{.0342} = -1.43$. Because -1.43 is not

≤ -1.895, do not reject H_0. The data do not contradict this assertion.

c. $\Sigma x_i = 306$, $\Sigma x_i^2 = 7946$, so $\displaystyle\sum(x_i - \bar{x})^2 = 7946 - \dfrac{(306)^2}{12} = 143$ here, as

contrasted with 182 for the given 9 x_i's. Even though the sample size for the proposed x values is larger, the original set of values is preferable.

d. $(t_{.025,7})(s)\sqrt{\dfrac{1}{9} + \dfrac{9(28 - 25.33)^2}{1638}} = (2.365)(.4608)(.3877) = .42$, and

$\hat{\beta}_0 + \hat{\beta}_1(28) = 10.02$, so the 95% CI is $10.02 \pm .42 = (9.60, 10.44)$.

77.

a. The plot suggests a strong linear relationship between x and y.

b. $n = 9, \Sigma x_i = 1797, \Sigma x_i^2 = 4334.41, \Sigma y_i = 7.28, \Sigma y_i^2 = 7.4028$ and

$\Sigma x_i y_i = 178.683$, so $\hat{\beta}_1 = \dfrac{299.931}{6717.6} = .04464854$, $\hat{\beta}_0 = -.08259353$, and the

equation of the estimated line is $\hat{y} = -.08259 + .044649x$.

c. $SSE = 7.4028 - (-601281) - 7.977935 = .026146$,

$SST = 7.4028 - \dfrac{(7.28)^2}{9} = 1.5141$, and $r^2 = 1 - \dfrac{SSE}{SST} = .983$, so 93.8% of the

observed variation can be explained by the model relationship.

d. $\hat{y}_4 = -.08259 - (.044649)(19.1) = .7702$, and $y_4 - \hat{y}_4 = .68 - .7702 = -.0902$.

e. $s = .06112$, and $s_{\hat{\beta}_1} = \dfrac{.06112}{\sqrt{746.4}} = .002237$, so the value of t for testing $H_0: \beta_1 = 0$

v. $H_a: \beta_1 \neq 0$ is $t = \dfrac{.044649}{.002237} = 19.96$. From Table A.5, $t_{.0005,7} = 5.408$, so the 2-

sided P-value is less than $2(.005) = .001$. There is strong evidence for a useful relationship.

f. A 95% CI for β_1 is $.044649 \pm (2.365)(.002237) = .044649 \pm .005291 = (.0394, .0499)$.

g. A 95% CI for $\beta_0 + \beta_1(20)$ is $.810 \pm (2.365)(.002237)(.3333356)$
$= .810 \pm .048 = (.762, .858)$

79. $SSE = \Sigma y^2 - \hat{\beta}_0 \Sigma y - \hat{\beta}_1 \Sigma xy$. Substituting $\hat{\beta}_0 = \dfrac{\Sigma y - \hat{\beta}_1 \Sigma x}{n}$, SSE becomes

$SSE = \Sigma y^2 - \dfrac{\Sigma y (\Sigma y - \hat{\beta}_1 \Sigma x)}{n} - \hat{\beta}_1 \Sigma xy = \Sigma y^2 - \dfrac{(\Sigma y)^2}{n} + \dfrac{\hat{\beta}_1 \Sigma x \Sigma y}{n} - \hat{\beta}_1 \Sigma xy$

$= \left[\Sigma y^2 - \dfrac{(\Sigma y)^2}{n} \right] - \hat{\beta}_1 \left[\Sigma xy - \dfrac{\Sigma x \Sigma y}{n} \right] = S_{yy} - \hat{\beta}_1 S_{xy}$, as desired.

81.

a. With $S_{xx} = \sum (x_i - \bar{x})^2$ and $S_{yy} = \sum (y_i - \bar{y})^2$, note that $\dfrac{s_y}{s_x} = \sqrt{\dfrac{S_{yy}}{S_{xx}}}$ (since the factor $n-1$ appears in both the numerator and denominator of the first ratio and, so, cancels). Thus

$$y = \hat{\beta}_0 + \hat{\beta}_1 x = \bar{y} + \hat{\beta}_1 (x - \bar{x}) = \bar{y} + \frac{S_{xy}}{S_{xx}} (x - \bar{x}) = \bar{y} + \sqrt{\frac{S_{yy}}{S_{xx}}} \cdot \frac{S_{xy}}{\sqrt{S_{xx} S_{yy}}} (x - \bar{x})$$

$$= \bar{y} + \frac{s_y}{s_x} \cdot r \cdot (x - \bar{x}), \text{ as desired.}$$

b. By .573 s.d.'s above, (above, since $r < 0$) or (since $s_y = 4.3143$) an amount 2.4721 above.

83. Using the notation of the exercise above, $SST = S_{yy}$, and $SSE = S_{yy} - \hat{\beta}_1 S_{xy}$

$$= S_{yy} - \frac{S_{xy}^2}{S_{xx}}, \text{ so } 1 - \frac{SSE}{SST} = 1 - \frac{S_{yy} - \dfrac{S_{xy}^2}{S_{xx}}}{S_{yy}} = \frac{S_{xy}^2}{S_{xx} S_{yy}} = r^2, \text{ as desired.}$$

85.   Using Minitab, we create a scatterplot to see if a linear regression model is appropriate.

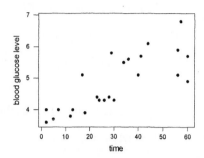

A linear model is reasonable; although it appears that the variance in y gets larger as x increases. The Minitab output follows:

```
The regression equation is
blood glucose level = 3.70 + 0.0379 time

Predictor          Coef         StDev            T         P
Constant         3.6965        0.2159        17.12     0.000
time            0.037895      0.006137         6.17     0.000

S = 0.5525      R-Sq = 63.4%      R-Sq(adj) = 61.7%

Analysis of Variance

Source             DF            SS            MS          F         P
Regression          1        11.638        11.638      38.12     0.000
Residual Error     22         6.716         0.305
Total              23        18.353
```

The coefficient of determination of 63.4% indicates that only a moderate percentage of the variation in y can be explained by the change in x. A test of model utility indicates that time is a significant predictor of blood glucose level. ($t = 6.17$, $p = 0.0$). A point estimate for blood glucose level when time = 30 minutes is 4.833%. We would expect the average blood glucose level at 30 minutes to be between 4.599 and 5.067, with 95% confidence.

87.   For the second boiler, $n = 6$, $\Sigma x_i = 125$, $\Sigma y_i = 472.0$, $\Sigma x_i^2 = 3625$,

$\Sigma y_i^2 = 37{,}140.82$, and $\Sigma x_i y_i = 9749.5$, giving $\hat{\gamma}_1 =$ estimated slope

$= \dfrac{-503}{6125} = -.0821224$, $\hat{\gamma}_0 = 80.377551$, $SSE_2 = 3.26827$, $SSx_2 = 1020.833$. For

boiler #1, $n = 8$, $\hat{\beta}_1 = -.1333$, $SSE_1 = 8.733$, and $SSx_1 = 1442.875$.

Thus, $\hat{\sigma}^2 = \dfrac{8.733 + 3.286}{10} = 1.2$, $\hat{\sigma} = 1.095$, and $t = \dfrac{-.1333 + .0821}{1.095\sqrt{\frac{1}{1442.875} + \frac{1}{1020.833}}}$

$= \dfrac{-.0512}{.0448} = -1.14$. Since $t_{.025,10} = 2.228$ and -1.14 is neither ≥ 2.228 nor ≤ -2.228,

H_0 is not rejected. It is plausible that $\beta_1 = \gamma_1$.

CHAPTER 13

Section 13.1

1.

    a. $\bar{x} = 15$ and $\sum (x_i - \bar{x})^2 = 250$, so SD of $Y_i - \hat{Y}_i$ is $10\sqrt{1 - \dfrac{1}{5} - \dfrac{(x_i - 15)^2}{250}} = 6.32$,
    8.37, 8.94, 8.37, and 6.32 for $i = 1, 2, 3, 4, 5$.

    b. Now $\bar{x} = 20$ and $\sum (x_i - \bar{x})^2 = 1250$, giving standard deviations 7.87, 8.49, 8.83,
    8.94, and 2.83 for $i = 1, 2, 3, 4, 5$.

    c. The deviation from the estimated line is likely to be much smaller for the
    observation made in the experiment of **b** for x = 50 than for the experiment of **a**
    when x = 25. That is, the observation (50, Y) is more likely to fall close to the
    least squares line than is (25, Y).

3.

    a. This plot indicates there are no outliers, the variance of ε is reasonably constant,
    and the ε's are normally distributed. A straight-line regression function is a
    reasonable choice for a model.

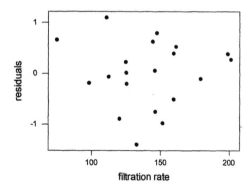

b. We need $S_{xx} = \sum (x_i - \bar{x})^2 = 415,914.85 - \dfrac{(2817.9)^2}{20} = 18,886.8295$. Then each e_i^*

can be calculated as follows: $e_i^* = \dfrac{e_i}{.4427\sqrt{1 + \dfrac{1}{20} + \dfrac{(x_i - 140.895)^2}{18,886.8295}}}$. The table

below shows the values:

standardized residuals	e/e_i^*	standardized residuals	e/e_i^*
−0.31064	0.644053	0.6175	0.64218
−0.30593	0.614697	0.09062	0.64802
0.4791	0.578669	1.16776	0.565003
1.2307	0.647714	−1.50205	0.646461
−1.15021	0.648002	0.96313	0.648257
0.34881	0.643706	0.019	0.643881
−0.09872	0.633428	0.65644	0.584858
−1.39034	0.640683	−2.1562	0.647182
0.82185	0.640975	−0.79038	0.642113
−0.15998	0.621857	1.73943	0.631795

Notice that if $e_i^* \approx e/s$, then $e/e_i^* \approx s$. All of the e/e_i^*'s range between .57 and .65, which are close to s.

c. This plot looks very much the same as the one in part a.

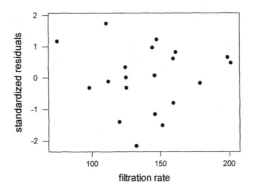

5.

   a. 97.7% of the variation in ice thickness can be explained by the linear relationship between it and elapsed time. Based on this value, it is tempting to assume an approximately linear relationship; however, r^2 does <u>not</u> measure the aptness of the linear model.

   b. The residual plot shows a curve in the data, suggesting a non–linear relationship exists. One observation (5.5, –3.14) is extreme.

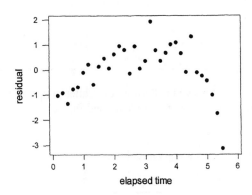

7.

a.

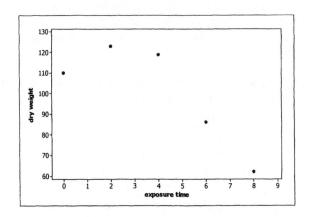

There is an obvious curved pattern in the scatter plot, which suggests that a simple linear model will not provide a good fit.

b. The $\hat{y}'s$, e's, and e^*'s are given below:

x	y	$\hat{y}$	e	e*
0	110	126.6	−16.6	−1.55
2	123	113.3	9.7	.68
4	119	100.0	19.0	1.25
6	86	86.7	−.7	−.05
8	62	73.4	−11.4	−1.06

Both residual plots (not shown) look similar to the original scatter plot and strongly suggest a quadratic (parabolic) relationship between x and y.

9. Both a scatter plot and residual plot (based on the simple linear regression model) for the first data set suggest that a simple linear regression model is reasonable, with no pattern or influential data points which would indicate that the model should be modified. However, scatter plots for the other three data sets reveal difficulties.

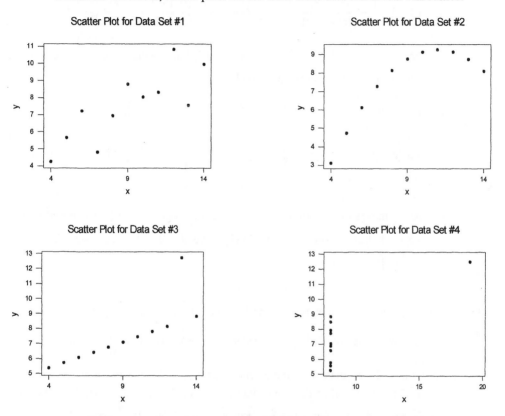

For data set #2, a quadratic function would clearly provide a much better fit. For data set #3, the relationship is perfectly linear except one outlier, which has obviously greatly influenced the fit even though its x value is not unusually large or small. One might investigate this observation to see whether it was mistyped and/or it merits deletion. For data set #4 it is clear that the slope of the least squares line has been determined entirely by the outlier, so this point is extremely influential. A linear model is completely inappropriate for data set #4; if anything, a modified logistic regression model (see Section 13.2) might be more appropriate.

11.

a. $Y_i - \hat{Y}_i = Y_i - \bar{Y} - \hat{\beta}_1(x_i - \bar{x}) = Y_i - \frac{1}{n}\sum_j Y_j - \dfrac{(x_i - \bar{x})\sum_j (x_j - \bar{x})Y_j}{\sum_j (x_j - \bar{x})^2} = \sum_j c_j Y_j$, where

$c_j = 1 - \dfrac{1}{n} - \dfrac{(x_i - \bar{x})^2}{n\sum (x_j - \bar{x})^2}$ for $j = i$ and $c_j = 1 - \dfrac{1}{n} - \dfrac{(x_i - \bar{x})(x_j - \bar{x})}{\sum (x_j - \bar{x})^2}$ for $j \neq i$.

Thus $Var(Y_i - \hat{Y}_i) = \sum Var(c_j Y_j)$ (since the Y_j's are independent) $= \sigma^2 \sum c_j^2$ which, after some algebra, gives equation (13.2).

b. $\sigma^2 = Var(Y_i) = Var(\hat{Y}_i + (Y_i - \hat{Y}_i)) = Var(\hat{Y}_i) + Var(Y_i - \hat{Y}_i)$, so

$Var(Y_i - \hat{Y}_i) = \sigma^2 - Var(\hat{Y}_i) = \sigma^2 - \sigma^2\left[\dfrac{1}{n} + \dfrac{(x_i - \bar{x})^2}{\sum (x_j - \bar{x})^2}\right]$, which is exactly (13.2).

c. As x_i moves further from $\bar{x}$, $(x_i - \bar{x})^2$ grows larger, so $Var(\hat{Y}_i)$ increases (since $(x_i - \bar{x})^2$ has a positive sign in $Var(\hat{Y}_i)$), but $Var(Y_i - \hat{Y}_i)$ decreases (since $(x_i - \bar{x})^2$ has a negative sign).

13. The distribution of any particular standardized residual is also a t distribution with df $= n–2$, since e_i^* is obtained by taking standard normal variable $\dfrac{Y_i - \hat{Y}_i}{\sigma_{Y_i - \hat{Y}}}$ and substituting the estimate of σ in the denominator (exactly as in the predicted value case). With E_i^* denoting the ith standardized residual as a random variable, when $n = 25$, E_i^* has a t distribution with 23 df. Since $t_{.01,23} = 2.50$, P(E_i^* outside (–2.50, 2.50)) $= P(E_i^* \geq 2.50) + P(E_i^* \leq -2.50) = .01 + .01 = .02$.

Section 13.2

15.

   a. The plot has a curved pattern. A linear model would not be appropriate.

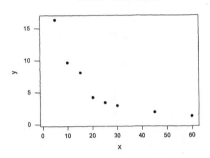

Scatter Plot of Y vs X

   b. In this plot we have a strong linear pattern.

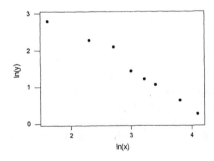

Scatter Plot of ln(Y) vs ln(X)

   c. The linear pattern in **b** above would indicate that a transformed regression using the natural log of both x and y would be appropriate. The probabilistic model is then $y = \alpha x^{\beta} \cdot \varepsilon$ (the power function with an error term).

d. A regression of ln(y) on ln(x) yields the equation $\ln(y) = 4.6384 - 1.04920 \ln(x)$. Using Minitab we can get a PI for y when x = 20 by first transforming the x value: $\ln(20) = 2.996$. The computer generated 95% P.I. for ln(y) when ln(x) = 2.996 is (1.1188, 1.8712). We must now take the antilog to return to the original units of Y: $\left(e^{1.1188}, e^{1.8712}\right) = (3.06, 6.50)$.

e. A computer generated residual analysis:

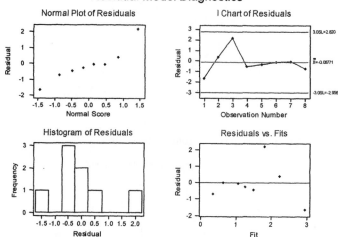

Residual Model Diagnostics

Looking at the residual vs. fits (bottom right), one standardized residual, corresponding to the third observation, is a bit large. There are only two positive standardized residuals, but two others are essentially 0. The patterns in the residual plot and the normal probability plot (upper left) are marginally acceptable.

17.

a. $\Sigma x_i' = 15.501$, $\Sigma y_i' = 13.352$, $\Sigma x_i'^2 = 20.228$, $\Sigma y_i'^2 = 16.572$, $\Sigma x_i' y_i' = 18.109$, from which $\hat{\beta}_1 = 1.254$ and $\hat{\beta}_0 = -.468$ so $\hat{\beta} = \hat{\beta}_1 = 1.254$ and $\hat{\alpha} = e^{-.468} = .626$.

b. The plots give strong support to this choice of model; in addition, $r^2 = .960$ for the transformed data.

c. SSE $= .11536$ (computer printout), $s = .1024$, and the estimated SD of $\hat{\beta}_1$ is

$.0775$, so $t = \dfrac{1.25 - 1.33}{.0775} = -1.07$. Since -1.07 is not $\le -t_{.05,11} = -1.796$, H_0

cannot be rejected in favor of H_a.

d. The claim that $\mu_{Y.5} = 2\mu_{Y.2.5}$ is equivalent to $\alpha \cdot 5^\beta = 2\alpha(2.5)^\beta$, or that $\beta = 1$.

Thus we wish test $H_0 : \beta = 1$ vs. $H_a : \beta \ne 1$. With $t = \dfrac{1 - 1.33}{.0775} = -4.30 \le$

$-t_{.005,11} \le -3.106$, H_0 is rejected at level $.01$.

19.

a. No, there is definite curvature in the plot.

b. With x = temperature and y = lifetime, a linear relationship between ln(lifetime)

and 1/temperature implies a model $y = \exp(\alpha + \beta/x + \varepsilon)$. Let $x' = \dfrac{1}{temp}$ and

$y' = \ln(lifetime)$. Plotting y' vs. x' gives a plot which has a pronounced linear

appearance (and in fact $r^2 = .954$ for the straight line fit).

c. $\Sigma x_i' = .082273$, $\Sigma y_i' = 123.64$, $\Sigma x_i'^2 = .00037813$, $\Sigma y_i'^2 = 879.88$,

$\Sigma x_i' \, y_i' = .57295$, from which $\hat{\beta} = 3735.4485$ and $\hat{\alpha} = -10.2045$ (values read

from computer output). With $x = 220$, $x' = .004545$ so

$\hat{y}' = -10.2045 + 3735.4485(.004545) = 6.7748$ and thus $\hat{y} = e^{\hat{y}} = 875.50$.

d. For the transformed data, SSE $= 1.39857$, and $n_1 = n_2 = n_3 = 6$, $\bar{y}_{1.}' = 8.44695$,

$\bar{y}_{2.}' = 6.83157$, $\bar{y}_{3.}' = 5.32891$, from which SSPE $= 1.36594$, SSLF $= .02993$,

$f = \dfrac{.02993/1}{1.36594/15} = .33$. Comparing this to $F_{.01,1,15} = 8.68$, it is clear that H_0

cannot be rejected.

21.

a. The suggested model is $Y = \beta_0 + \beta_1(x') + \varepsilon$ where $x' = \dfrac{10^4}{x}$. The summary

quantities are $\Sigma x_i' = 159.01$, $\Sigma y_i = 121.50$, $\Sigma x_i'^2 = 4058.8$, $\Sigma y_i^2 = 1865.2$,

$\Sigma x_i' \, y_i = 2281.6$, from which $\hat{\beta}_1 = -.1485$ and $\hat{\beta}_0 = 18.1391$, and the estimated

regression function is $y = 18.1391 - \dfrac{1485}{x}$.

b. $x = 500 \Rightarrow \hat{y} = 18.1391 - \dfrac{1485}{500} = 15.17$.

23. $Var(Y) = Var(\alpha e^{\beta x} \cdot \varepsilon) = \left[\alpha e^{\beta x}\right]^2 \cdot Var(\varepsilon) = \alpha^2 e^{2\beta x} \cdot \tau^2$, where we have set

$Var(\varepsilon) = \tau^2$. If $\beta > 0$, this is an increasing function of x so we expect more spread in
y for large x than for small x, while the situation is reversed if $\beta < 0$. It is important to
realize that a scatter plot of data generated from this model will not spread out
uniformly about the exponential regression function throughout the range of x values;
the spread will only be uniform on the transformed scale. Similar results hold for the
multiplicative power model.

25.   The point estimate of β_1 is $\hat{\beta}_1 = .17772$, so the estimate of the odds ratio is

$e^{\hat{\beta}_1} = e^{.17772} \approx 1.194$. That is, when the amount of experience increases by one year
(i.e. a one unit increase in x), we estimate that the odds increase by about 1.194. The
z value of 2.70 and its corresponding P-value of .007 imply that the null hypothesis
$H_0: \beta_1 = 0$ can be rejected at any reasonable significance level. Therefore, there is
clear evidence that β_1 is not zero, which means that experience does appear to affect
the likelihood of successfully performing the task. This is consistent with the
confidence interval (1.05, 1.36) for the odds ratio given in the printout, since this
interval does not contain the value 1. A graph of $\hat{\pi}$ appears below.

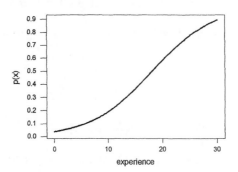

Section 13.3

27.

a. A scatter plot of the data indicated a quadratic regression model might be appropriate.

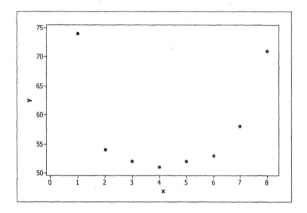

b. $\hat{y} = 84.482 - 15.875(6) + 1.7679(6)^2 = 52.88$; residual = $y_6 - \hat{y}_6 = 53 - 52.88 = .12$

c. $SST = \Sigma y_i^2 - \dfrac{(\Sigma y_i)^2}{n} = 586.88$, so $R^2 = 1 - \dfrac{61.77}{586.88} = .895$.

d. None of the standardized residuals exceeds 2 in magnitude, suggesting none of the observations are outliers. The ordered z percentiles needed for the normal probability plot are $-1.53, -.89, -.49, -.16, .16, .49, .89$, and 1.53. The normal probability plot below does not exhibit any troublesome features.

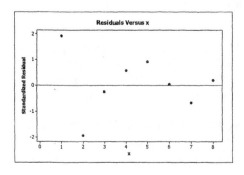
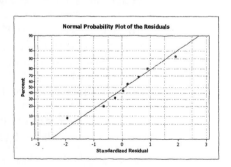

e. $\hat{\mu}_{Y\cdot 6} = 52.88$ (from b) and $t_{.025,n-3} = t_{.025,5} = 2.571$, so the CI is
$52.88 \pm (2.571)(1.69) = 52.88 \pm 4.34 = (48.54, 57.22)$.

f. $SSE = 61.77$, so $s^2 = \dfrac{61.77}{5} = 12.35$ and $s\{\text{pred}\}\ \sqrt{12.35 + (1.69)^2} = 3.90$.
The PI is $52.88 \pm (2.571)(3.90) = 52.88 \pm 10.03 = (42.85, 62.91)$.

29.

a. The table below displays the y-values, fits, and residuals. From this, $SSE = \sum e^2 = 16.8$, $s^2 = SSE/(n-3) = 4.2$, and $s = 2.05$.

y	$\hat{y}$	$e = y - \hat{y}$
81	82.1342	−1.13420
83	80.7771	2.22292
79	79.8502	−0.85022
75	72.8583	2.14174
70	72.1567	−2.15670
43	43.6398	−0.63985
22	21.5837	0.41630

b. $SST = \sum (y - \bar{y})^2 = \sum (y - 64.71)^2 = 3233.4$, so $R^2 = 1 - SSE/SST = 1 - 16.8/3233.4 = .995$, or 99.5%. 995% of the variation in free–flow can be explained by the quadratic regression relationship with viscosity.

c. We want to test the hypotheses H_0: $\beta_2 = 0$ v. H_a: $\beta_2 \neq 0$. Assuming all inference assumptions are met, the relevant t statistic is $t = \dfrac{-.0031662 - 0}{.0004835} = -6.55$. At df = $n - 3 = 4$, the corresponding P–value is $2P(|t| > 6.55) < .004$. At any reasonable significance level, we would reject H_0 and conclude that the quadratic predictor indeed belongs in the regression model.

d. Two intervals with at least 95% simultaneous confidence requires individual confidence equal to $100\% - 5\%/2 = 97.5\%$. To use the t–table, round up to 98%: $t_{.01,4} = 3.747$. The two confidence intervals are $2.1885 \pm 3.747(.4050) = (.671, 3.706)$ for β_1 and $-.0031662 \pm 3.747(.0004835) = (-.00498, -.00135)$ for β_2. [In fact, we are at least 96% confident β_1 and β_2 lie in these intervals.]

e. Plug into the regression equation to get $\hat{y} = 72.858$. Then a 95% CI for $\mu_{Y\cdot 400}$ is $72.858 \pm 3.747(1.198) = (69.531, 76.186)$. For the PI, $s\{\text{pred}\} = \sqrt{s^2 + s_{\hat{Y}}^2} = \sqrt{4.2 + (1.198)^2} = 2.374$, so a 95% PI for Y when $x = 400$ is $72.858 \pm 3.747(2.374) = (66.271, 79.446)$.

31.

a. Using Minitab, the regression equation is $\hat{y} = 13.6 + 11.4x - 1.72x^2$.

b. Again, using Minitab, the predicted and residual values are:

$\hat{y}$:	23.327	23.327	29.587	31.814	31.814	31.814	20.317
$y - \hat{y}$:	−.327	1.173	1.587	.914	.186	1.786	−.317

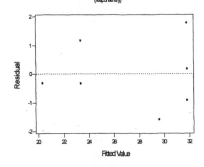

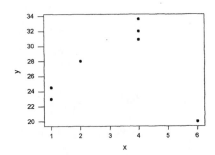

The residual plot is consistent with a quadratic model (no pattern which would suggest modification), but it is clear from the scatter plot that the point (6, 20) has had a great influence on the fit – it is the point which forced the fitted quadratic to have a maximum between 3 and 4 rather than, for example, continuing to curve slowly upward to a maximum someplace to the right of x = 6.

c. From Minitab output, $s^2 = MSE = 2.040$, and $R^2 = 94.7\%$. The quadratic model thus explains 94.7% of the variation in the observed y's, which suggests that the model fits the data quite well.

d. $\sigma^2 = Var(\hat{Y}_i) + Var\left(Y_i - \hat{Y}_i\right)$ suggests that we can estimate $Var\left(Y_i - \hat{Y}_i\right)$ by $s^2 - s_{\hat{y}}^2$ and then take the square root to obtain the estimated standard deviation of each residual. This gives $\sqrt{2.040 - (.955)^2} = 1.059$, (and similarly for all points) 10.59, 1.236, 1.196, 1.196, 1.196, and .233 as the estimated SD's of the residuals. The standardized residuals are then computed as $\dfrac{-.327}{1.059} = -.31$, (and similarly) 1.10, −1.28, −.76, .16, 1.49, and −1.28, none of which are unusually large. (Note: Minitab regression output can produce these values.) The resulting residual plot is virtually identical to the plot of **b**. $\dfrac{y - \hat{y}}{s} = \dfrac{-.327}{1.426} = -.229 \neq -.31$, so standardizing using just s would not yield the correct standardized residuals.

Chapter 13: Nonlinear and Multiple Regression

e. $Var(Y_f) + Var(\hat{Y}_f)$ is estimated by $2.040 + (.777)^2 = 2.638$, so

$s_{y_f + \hat{y}_f} = \sqrt{2.638} = 1.624$. With $\hat{y} = 31.81$ and $t_{.05,4} = 2.132$, the desired PI is

$31.81 \pm (2.132)(1.624) = (28.35, 35.27)$.

33.

a. $\bar{x} = 20$ and $s_x = 10.8012$ so $x' = \dfrac{x - 20}{10.8012}$. For $x = 20$, $x' = 0$, and

$\hat{y} = \hat{\beta}_0^* = .9671$. For $x = 25$, $x' = .4629$, so

$\hat{y} = .9671 - .0502(.4629) - .0176(.4629)^2 + .0062(.4629)^3 = .9407$.

b. $\hat{y} = .9671 - .0502\left(\dfrac{x - 20}{10.8012}\right) - .0176\left(\dfrac{x - 20}{10.8012}\right)^2 + .0062\left(\dfrac{x - 20}{10.8012}\right)^3$

$.00000492x^3 - .000446058x^2 + .007290688x + .96034944$.

c. $t = \dfrac{.0062}{.0031} = 2.00$. We reject H_0 if either $t \geq t_{.025, n-4} = t_{.025, 3} = 3.182$ or if

$t \leq -3.182$. Since 2.00 is neither ≥ 3.182 nor ≤ -3.182, we cannot reject H_0; the cubic term should be deleted.

d. $SSE = \Sigma(y_i - \hat{y}_i)$ and the $\hat{y}_i$'s are the same from the standardized as from the unstandardized model, so SSE, SST, and R^2 will be identical for the two models.

e. $\Sigma y_i^2 = 6.355538$, $\Sigma y_i = 6.664$, so $SST = .011410$. For the quadratic model $R^2 = .987$ and for the cubic model, $R^2 = .994$; the two R^2 values are very close, suggesting intuitively that the cubic term is relatively unimportant.

35. $Y' = \ln(Y) = \ln \alpha + \beta x + \gamma x^2 + \ln(\varepsilon) = \beta_0 + \beta_1 x + \beta_2 x^2 + \varepsilon'$ where $\varepsilon' = \ln(\varepsilon)$,
$\beta_0 = \ln(\alpha)$, $\beta_1 = \beta$, and $\beta_2 = \gamma$. That is, we should fit a quadratic to $(x, \ln(y))$.
The resulting estimated quadratic (from computer output) is
$2.00397 + .1799x - .0022x^2$, so $\hat{\beta} = .1799$, $\hat{\gamma} = -.0022$, and $\hat{\alpha} = e^{2.0397} = 7.6883$.
(The $\ln(y)$'s are 3.6136, 4.2499, 4.6977, 5.1773, and 5.4189, and the summary quantities can then be computed as before.)

Section 13.4

37.

    a. The mean value of y when $x_1 = 50$ and $x_2 = 3$ is

        $\mu_{y \cdot 50,3} = -.800 + .060(50) + .900(3) = 4.9$ hours.

    b. When the number of deliveries (x_2) is held fixed, then average change in travel time associated with a one–mile (i.e. one unit) increase in distance traveled (x_1) is .060 hours. Similarly, when distance traveled (x_1) is held fixed, then the average change in travel time associated with on extra delivery (i.e., a one unit increase in x_2) is .900 hours.

    c. Under the assumption that y follows a normal distribution, the mean and standard deviation of this distribution are 4.9 (because $x_1 = 50$ and $x_2 = 3$) and $\sigma = .5$ (since the standard deviation is assumed to be constant regardless of the values of x_1 and x_2). Therefore, $P(y \le 6) = P\left(z \le \dfrac{6 - 4.9}{.5} \right) = P(z \le 2.20) = .9861$. That is,

    in the long run, about 98.6% of all days will result in a travel time of at most 6 hours.

39.

    a. For $x_1 = 2$, $x_2 = 8$ (remember the units of x_2 are in 1000's) and $x_3 = 1$ (since the outlet has a drive–up window) the average sales are

        $\hat{y} = 10.00 - 1.2(2) + 6.8(8) + 15.3(1) = 77.3$ (i.e., \$77,300).

    b. For $x_1 = 3$, $x_2 = 5$, and $x_3 = 0$ the average sales are

        $\hat{y} = 10.00 - 1.2(3) + 6.8(5) + 15.3(0) = 40.4$ (i.e., \$40,400).

    c. When the number of competing outlets (x_1) and the number of people within a one-mile radius (x_2) remain fixed, the expected sales will increase by \$15,300 when an outlet has a drive-up window.

41. $H_0 : \beta_1 = \beta_2 = \ldots = \beta_6 = 0$ vs. H_a: at least one among $\beta_1, \ldots, \beta_6$ is not zero. The test

statistic is $F = \dfrac{R^2 / k}{(1 - R^2)/(n - k - 1)}$. H_0 will be rejected if $f \ge F_{.05,6,30} = 2.42$.

$f = \dfrac{.83 / 6}{(1 - .83)/30} = 24.41$. Because $24.41 \ge 2.42$, H_0 is rejected and the model is judged useful.

43.

    a. The coefficient of multiple determination is $R^2 = 78\%$. So 78% of the observed variation in surface area can be attributed to the stated approximate relationship between surface area and the predictors.

    b. $x_1 = 2.6$, $x_2 = 250$, and $x_1 x_2 = (2.6)(250) = 650$, so
$$\hat{y} = 185.49 - 45.97(2.6) - 0.3015(250) + 0.0888(650) = 48.313$$

    c. No, it is not legitimate to interpret β_1 in this way. It is not possible to increase by 1 unit the cobalt content, x_1, while keeping the interaction predictor, x_3, fixed. When x_1 changes, so does x_3, since $x_3 = x_1 x_2$.

    d. Yes, there appears to be a useful linear relationship between y and the predictors. We determine this by observing that the P-value corresponding to the model utility test is $< .0001$ (F test statistic $= 18.924$).

    e. We wish to test $H_0 : \beta_3 = 0$ v. $H_a : \beta_3 \neq 0$. The test statistic is $t = 3.496$, with a corresponding P-value of .0030. Since P-value $< \alpha = .01$, we reject H_0 and conclude that the interaction predictor does provide useful information about y.

    f. A 95% C.I. for the mean value of surface area under the stated circumstances requires the following quantities: $t_{.025,16} = 2.120$, and
$$\hat{y} = 185.49 - 45.97(2) - 0.3015(500) + 0.0888(2)(500) = 31.598.$$ So the 95% CI is
$$31.598 \pm (2.120)(4.69) = 31.598 \pm 9.9428 = (21.6552, 41.5408).$$

45.

    a. The hypotheses are $H_0: \beta_1 = \beta_2 = \beta_3 = \beta_4 = 0$ vs. H_a: at least one $\beta_i \neq 0$. The test statistic is $f = \dfrac{R^2 / k}{(1 - R^2)/(n - k - 1)} = \dfrac{.946 / 4}{(1 - .946)/20} = 87.6 \geq F_{.001,4,20} = 7.10$ (the smallest available significance level from Table A.9), so we can reject H_0 at any significance level. We conclude that at least one of the four predictor variables appears to provide useful information about tenacity.

    b. The adjusted R^2 value is $1 - \dfrac{n-1}{n - (k+1)} \left(\dfrac{SSE}{SST} \right) = 1 - \dfrac{n-1}{n - (k+1)} \left(1 - R^2 \right)$
$$= 1 - \frac{24}{20}(1 - .946) = .935,$$ which does not differ much from $R^2 = .946$.

244

c. The estimated average tenacity when $x_1 = 16.5$, $x_2 = 50$, $x_3 = 3$, and $x_4 = 5$ is
$\hat{y} = 6.121 - .082(16.5) + .113(50) + .256(3) - .219(5) = 10.091$. For a 99% C.I.,
$t_{.005,20} = 2.845$, so the interval is $10.091 \pm 2.845(.350) = (9.095, 11.087)$.
Therefore, when the four predictors are as specified in this problem, the true average tenacity is estimated to be between 9.095 and 11.087.

47.

a. For a 1% increase in the percentage plastics, we would expect a 28.9 kcal/kg increase in energy content. Also, for a 1% increase in the moisture, we would expect a 37.4 kcal/kg decrease in energy content.

b. The hypotheses are H_0: $\beta_1 = \beta_2 = \beta_3 = \beta_4 = 0$ vs. H_a: at least one $\beta_i \neq 0$. The value of the F test statistic is 167.71, with a corresponding p–value that is extremely small. So, we reject H_0 and conclude that at least one of the four predictors is useful in predicting energy content, using a linear model.

c. H_0: $\beta_3 = 0$ vs. H_a: $\beta_3 \neq 0$. The value of the t test statistic is $t = 2.24$, with a corresponding P-value of .034, which is less than the significance level of .05. So we can reject H_0 and conclude that percentage garbage provides useful information about energy consumption, given that the other three predictors remain in the model.

d. $\hat{y} = 2244.9 + 28.925(20) + 7.644(25) + 4.297(40) - 37.354(45) = 1505.5$, and $t_{.025,25} = 2.060$. A 95% C.I for the true average energy content under these circumstances is $1505.5 \pm (2.060)(12.47) = 1505.5 \pm 25.69 = (1479.8, 1531.1)$. Because the interval is reasonably narrow, we would conclude that the mean energy content has been precisely estimated.

e. A 95% prediction interval for the energy content of a waste sample having the specified characteristics is $1505.5 \pm (2.060)\sqrt{(31.48)^2 + (12.47)^2}$
$= 1505.5 \pm 69.75 = (1435.7, 1575.2)$.

49.

a. $\hat{\mu}_{y\cdot18.9,43} = 21.967 = 96.8303$, and residual $= 91 - 96.8303 = -5.8303$.

b. $H_0: \beta_1 = \beta_2 = 0$ vs. H_a: at least one $\beta_i \neq 0; f = \dfrac{R^2 / k}{(1 - R^2)/(n - k - 1)} =$

$\dfrac{.798/2}{(1-.798)/9} = 14.90 \geq F_{.05,2,9} = 8.02$, so we reject H_0. The model appears useful.

c. $96.8303 \pm (2.262)(8.20) = (78.28, 115.38)$

d. $96.8303 \pm (2.262)\sqrt{24.45^2 + 8.20^2} = (38.50, 155.16)$

e. We find the center of the given 95% interval, 93.875, and half of the width, 57.845. This latter value is equal to $t_{.025,9}(s_{\hat{y}}) = 2.262(s_{\hat{y}})$, so $s_{\hat{y}} = 25.5725$. Then the 90% interval is $93.785 \pm (1.833)(25.5725) = (46.911, 140.659)$

f. With the P-value for $H_a: \beta_1 \neq 0$ being 0.208 (from given output), we would fail to reject H_0. This factor is not significant given x_2 is in the model.

g. With $R_k^2 = .768$ (full model) and $R_l^2 = .721$ (reduced model), we can use an alternative F statistic: $F = \dfrac{(R_k^2 - R_l^2)/(k - l)}{(1 - R_k^2)/(n - k - 1)}$. With $n = 12$, $k = 2$ and $l = 1$, we have $f = \dfrac{(.768 - .721)/(2 - 1)}{(1 - .768)/12 - 2 - 1} = 1.83$. $t^2 = (-1.36)^2 = 1.85$. The discrepancy can be attributed to rounding error.

51.

a. No, there is no pattern in the plots which would indicate that a transformation or the inclusion of other terms in the model would produce a substantially better fit.

b. $k = 5$, $n - (k+1) = 8$, so $H_0: \beta_1 = \ldots = \beta_5 = 0$ will be rejected if $f \geq F_{.05,5,8} = 3.69$;

$$f = \frac{.759/5}{.241/8} = 5.04 \geq 3.69, \text{ so we reject } H_0. \text{ At least one of the coefficients is not}$$

equal to zero.

c. When $x_1 = 8.0$ and $x_2 = 33.1$ the residual is $e = 2.71$ and the standardized residual is $e^* = .44$; since $e^* = e/(\text{SD of the residual})$, SD of residual $= e/e^* = 6.16$. Thus the estimated variance of $\hat{Y}$ is $(6.99)^2 - (6.16)^2 = 10.915$, so the estimated SD is 3.304. Since $\hat{y} = 24.29$ and $t_{.025,8} = 2.306$, the desired CI is

$$24.29 \pm 2.306(3.304) = (16.67, 31.91).$$

d. $F_{.05,3,8} = 4.07$, so $H_0 : \beta_3 = \beta_4 = \beta_5 = 0$ will be rejected if $f \geq 4.07$. With

$$SSE_k = (14-6)s^2 = 390.88, \ f = \frac{(894.95 - 390.88)/3}{390.88/8} = 3.44. \text{ Since } 3.44 < 4.07,$$

H_0 cannot be rejected and the quadratic terms should all be deleted. (N.B.: this is not a modification which would be suggested by a residual plot.)

53. Some possible questions might be:
   (1) Is this model useful in predicting deposition of poly-aromatic hydrocarbons? A test of model utility gives us an F = 84.39, with a P-value of 0.000. Thus, the model is useful.
   (2) Is x_1 a significant predictor of y in the presence of x_2? A test of $H_0: \beta_1 = 0$ v. $H_a: \beta_1 \neq 0$ gives us a t = 6.98 with a P-value of 0.000., so this predictor is significant.
   (3) A similar question, and solution for testing x_2 as a predictor yields a similar conclusion: With a P-value of 0.046, we would accept this predictor as significant if our significance level were anything larger than 0.046.

Section 13.5

55.

a. $\ln(Q) = Y = \ln(\alpha) + \beta \ln(a) + \gamma \ln(b) + \ln(\varepsilon) = \beta_0 + \beta_1 x_1 + \beta_2 x_2 + \varepsilon'$ where
$x_1 = \ln(a), x_2 = \ln(b), \beta_0 = \ln(\alpha), \beta_1 = \beta, \beta_2 = \gamma$ and $\varepsilon' = \ln(\varepsilon)$. Thus we
transform to $(y, x_1, x_2) = (\ln(Q), \ln(a), \ln(b))$ (take the natural log of the values of
each variable) and do a multiple linear regression. A computer analysis gave
$\hat{\beta}_0 = 1.5652$, $\hat{\beta}_1 = .9450$, and $\hat{\beta}_2 = .1815$. For $a = 10$ and $b = .01$, $x_1 = \ln(10)$
$= 2.3026$ and $x_2 = \ln(.01) = -4.6052$, from which $\hat{y} = 2.9053$ and
$\hat{Q} = e^{2.9053} = 18.27$.

b. Again taking the natural log, $Y = \ln(Q) = \ln(\alpha) + \beta a + \gamma b + \ln(\varepsilon)$, so to fit this
model it is necessary to take the natural log of each Q value (and not transform a
or b) before using multiple regression analysis.

c. We simply exponentiate each endpoint: $\left(e^{.217}, e^{1.755} \right) = (1.24, 5.78)$.

57.

k	R^2	Adj. R^2	$C_k = \dfrac{SSE_k}{s^2} + 2(k+1) - n$
1	.676	.647	138.2
2	.979	.975	2.7
3	.9819	.976	3.2
4	.9824		4

Where $s^2 = 5.9825$

a. Clearly the model with k = 2 is recommended on all counts.

b. No. Forward selection would let x_4 enter first and would not delete it at the next
stage.

59. The choice of a "best" model seems reasonably clear-cut. The model with 4 variables including all but the summerwood fiber variable would seem best. R^2 is as large as any of the models, including the 5-variable model. R^2 adjusted is at its maximum and CP is at its minimum. As a second choice, one might consider the model with $k = 3$ which excludes the summerwood fiber and springwood % variables.

61. If multicollinearity were present, at least one of the four R^2 values would be very close to 1, which is not the case. Therefore, we conclude that multicollinearity is not a problem in this data.

63. Before removing any observations, we should investigate their source (e.g., were measurements on that observation misread?) and their impact on the regression. To begin, Observation #7 deviates significantly from the pattern of the rest of the data (standardized residual $= -2.62$); if there's concern the PAH deposition was not measured properly, we might consider removing that point to improve the overall fit. If the observation was not mis-recorded, we should not remove the point.

We should also investigate Observation #6: Minitab gives $h_{66} = .846 > 3(2+1)/17$, indicating this observation has very high leverage. However, the standardized residual for #6 is not large, suggesting that it follows the regression pattern specified by the other observations. Its "influence" only comes from having a comparatively large x_1 value.

Supplementary Exercises

65.

a.

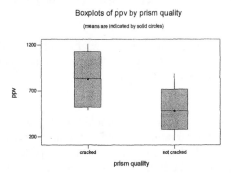

Boxplots of ppv by prism quality

(means are indicated by solid circles)

A two-sample t confidence interval, generated by Minitab:

```
prism qu      N     Mean     StDev   SE Mean
cracked      12      827       295        85
not cracke   18      483       234        55

95% CI for mu (cracked   ) - mu (not cracke): ( 132,   557)
```

b. The simple linear regression results in a significant model, r^2 is .577, but we have an extreme observation, with std resid = −4.11. Minitab output is below. Also run, but not included here was a model with an indicator for cracked/ not cracked, and for a model with the indicator and an interaction term. Neither improved the fit significantly.

```
The regression equation is
ratio = 1.00 -0.000018 ppv

Predictor          Coef        StDev            T          P
Constant        1.00161      0.00204       491.18      0.000
ppv          -0.00001827  0.00000295        -6.19      0.000

S = 0.004892     R-Sq = 57.7%      R-Sq(adj) = 56.2%

Analysis of Variance

Source            DF           SS           MS          F          P
Regression         1   0.00091571   0.00091571      38.26      0.000
Residual Error    28   0.00067016   0.00002393
Total             29   0.00158587

Unusual Observations
Obs         ppv        ratio          Fit  StDev Fit  Residual  St Resid
 29        1144     0.962000     0.980704   0.001786 -0.018704     -4.11R
R denotes an observation with a large standardized residual
```

67.

    a.   After accounting for all the other variables in the regression, we would expect the VO_2max to decrease by .0996, on average for each one–minute increase in the 1–mile walk time.

    b.   After accounting for all the other variables in the regression, we expect males to have a VO_2max that is .6566 L/min higher than females, on average.

    c. $\hat{y} = 3.5959 + .6566(1) + .0096(170) - .0996(11) - .0880(140) = 3.67$. The residual is $\hat{y} = (3.15 - 3.67) = -.52$.

    d. $R^2 = 1 - \dfrac{SSE}{SST} = 1 - \dfrac{30.1033}{102.3922} = .706$, or 70.6% of the observed variation in VO_2max can be attributed to the model relationship.

    e. $H_0 : \beta_1 = \beta_2 = \beta_3 = \beta_4 = 0$ will be rejected in favor of H_a: at least one among $\beta_1, \ldots, \beta_4 \neq 0$, if $f \geq F_{.05, 4, 15} = 8.25$. With $f = \dfrac{.706/4}{(1 - .706)/15} = 9.005 \geq 8.25$, so H_0 is rejected. It appears that the model specifies a useful relationship between VO_2max and at least one of the other predictors.

69.

    a.   Based on a scatter plot (below), a simple linear regression model would not be appropriate. Because of the slight, but obvious curvature, a quadratic model would probably be more appropriate.

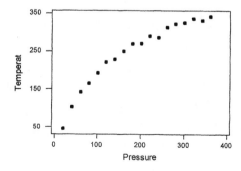

b. Using a quadratic model, a Minitab generated regression equation is $\hat{y} = 35.423 + 1.7191x - .0024753x^2$, and a point estimate of temperature when pressure is 200 is $\hat{y} = 280.23$. Minitab will also generate a 95% prediction interval of $(256.25, 304.22)$. That is, we are confident that when pressure is 200 psi, a single value of temperature will be between 256.25 and 304.22°F.

71.

a. Using Minitab to generate the first order regression model, we test the model utility (to see if any of the predictors are useful), and with $f = 21.03$ and a P-value of .000, we determine that at least one of the predictors is useful in predicting palladium content. Looking at the individual predictors, the p–value associated with the pH predictor has value .169, which would indicate that this predictor is unimportant in the presence of the others.

b. Testing $H_0 : \beta_1 = \ldots = \beta_{20} = 0$ vs. H_a: at least one of the β_i's $\neq 0$. With calculated statistic $f = 6.29$, and P-value .002, this model is also useful at any reasonable significance level.

c. Testing $H_0 : \beta_6 = \ldots = \beta_{20} = 0$ vs. H_a: at least one of the listed β_i's $\neq 0$, the test statistic is $f = \dfrac{(716.10 - 290.27)/(20 - 5)}{290.27(32 - 20 - 1)} = 1.07$. Using significance level .05, the rejection region would be $f \geq F_{.05,15,11} = 2.72$. Since $1.07 < 2.72$, we fail to reject H_0 and conclude that all the quadratic and interaction terms should not be included in the model. They do not add enough information to make this model significantly better than the simple first order model.

d. Partial output from Minitab follows, which shows all predictors as significant at level .05:

```
The regression equation is
pdconc = - 305 + 0.405 niconc + 69.3 pH - 0.161 temp + 0.993
currdens
          + 0.355 pallcont - 4.14 pHsq

Predictor        Coef        StDev         T         P
Constant      -304.85        93.98     -3.24     0.003
niconc         0.40484      0.09432      4.29     0.000
pH            69.27         21.96        3.15     0.004
temp          -0.16134      0.07055     -2.29     0.031
currdens       0.9929       0.3570       2.78     0.010
pallcont       0.35460      0.03381     10.49     0.000
pHsq          -4.138        1.293       -3.20     0.004
```

Chapter 13: Nonlinear and Multiple Regression

73.

a. We wish to test $H_0 : \beta_1 = \beta_2 = 0$ vs. H_a: either β_1 or $\beta_2 \neq 0$. The test statistic is

$$f = \frac{R^2 / k}{(1 - R^2)/(n - k - 1)} ,$$ where $k = 2$ for the quadratic model. The rejection region

is $f \geq F_{\alpha,k,n-k-1} = F_{.01,2,5} = 13.27$. $R^2 = 1 - \frac{.29}{202.88} = .9986$, giving $f = 1783 \gg$

13.27. The quadratic model is clearly useful.

b. The relevant hypotheses are $H_0 : \beta_2 = 0$ v. H_a: $\beta_2 \neq 0$. H_0 will be rejected at
level .001 if either $t \geq 6.869$ or $t \leq -6.869$ (df $= n - 3 = 5$).The test statistic value

is $t = \frac{\hat{\beta}_2}{s_{\hat{\beta}_2}} = \frac{-.00163141}{.00003391} = -48.1 \leq -6.869$, H_0 is rejected. The quadratic

predictor should be retained.

c. No. R^2 is extremely high for the quadratic model, so the marginal benefit of
including the cubic predictor would be essentially nil – and a scatter plot doesn't
show the type of curvature associated with a cubic model.

d. $t_{.025,5} = 2.571$, and $\hat{\beta}_0 + \hat{\beta}_1(100) + \hat{\beta}_2(100)^2 = 21.36$, so the CI is $21.36 \pm$
$2.571(.1141) = 21.36 \pm .29 = (21.07, 21.65)$.

e. First, we need to figure out s^2 based on the information we have been given.
$s^2 = MSE = \frac{SSE}{df} = \frac{.29}{5} = .058$. Then, the 95% P.I. is

$$21.36 \pm 2.571\sqrt{.058 + (.1141)^2} = 21.36 \pm 0.685 = (20.675, 22.045).$$

75.

a. $H_0 : \beta_1 = \beta_2 = 0$ will be rejected in favor of H_a: either β_1 or $\beta_2 \neq 0$ if f

$$= \frac{R^2 / k}{(1 - R^2)/(n - k - 1)} \geq F_{.01,2,7} = 9.55 . \quad SST = \Sigma y^2 - \frac{(\Sigma y)}{n} = 264.5 \text{ , so}$$

$R^2 = 1 - \frac{26.98}{264.5} = .898$, and $f = \frac{.898/2}{.102/7} = 30.8$. Because $30.8 \geq 9.55$, H_0 is rejected

at significance level .01 and the quadratic model is judged useful.

Chapter 13: Nonlinear and Multiple Regression

b. The hypotheses are $H_0 : \beta_2 = 0$ v. $H_a: \beta_2 \neq 0$. The test statistic value is

$$t = \frac{\hat{\beta}_2}{s_{\hat{\beta}_2}} = \frac{-2.3621}{.3073} = -7.69 \text{ , and } t_{.0005,7} = 5.408 \text{, so } H_0 \text{ is rejected at level .001. The}$$

quadratic predictor should not be eliminated.

c. $x = 1$ here, and $\hat{\mu}_{Y\cdot1} = \hat{\beta}_0 + \hat{\beta}_1(1) + \hat{\beta}_2(1)^2 = 45.96$. $t_{.025,7} = 1.895$, giving the CI
$45.96 \pm (1.895)(1.031) = (44.01, 47.91)$.

77.

a. estimate $= \hat{\beta}_0 + \hat{\beta}_1(15) + \hat{\beta}_2(3.5) = 180 + (1)(15) + (10.5)(3.5) = 231.75$

b. $R^2 = 1 - \dfrac{117.4}{1210.30} = .903$

c. $H_0 : \beta_1 = \beta_2 = 0$ v. H_a: either β_1 or $\beta_2 \neq 0$. $f = \dfrac{.903/2}{.097/9} = 41.9$, which greatly
exceeds $F_{.01,2,9}$ so there appears to be a useful linear relationship.

d. $s^2 = \dfrac{117.40}{12-3} = 13.044$, s{pred} $= \sqrt{s^2 + s_{\hat{Y}}^2} = 3.806$, and $t_{.025,9} = 2.262$. Hence,
the PI is $229.5 \pm 2.262(3.806) = (220.9, 238.1)$.

79. There are obviously several reasonable choices in each case. In **a**, the model with 6 carriers is a defensible choice on all three grounds, as are those with 7 and 8 carriers. The models with 7, 8, or 9 carriers in **b** merit serious consideration. These models merit consideration because R_k^2, MSE_k, and C_k meet the variable selection criteria given in Section 13.5.

Chapter 13: Nonlinear and Multiple Regression

81.

a. The relevant hypotheses are $H_0 : \beta_1 = \ldots = \beta_5 = 0$ vs. H_a: at least one among $\beta_1, \ldots, \beta_5 \neq 0$. $f = \dfrac{.827/5}{.173/11} = 106.1 \geq F_{.05,5,111} \approx 2.29$. Hence, H_0 is rejected in favor of the conclusion that there is a useful linear relationship between Y and at least one of the predictors.

b. $t_{.05,111} = 1.66$, so the CI is $.041 \pm (1.66)(.016) = .041 \pm .027 = (.014, .068)$. β_1 is the expected change in mortality rate associated with a one-unit increase in the particle reading when the other four predictors are held fixed; we can be 90% confident that $.014 < \beta_1 < .068$.

c. $H_0 : \beta_4 = 0$ will be rejected in favor of $H_a : \beta_4 \neq 0$ if $t = \dfrac{\hat{\beta}_4}{s_{\hat{\beta}_4}}$ is either ≥ 2.62 or ≤ -2.62. $t = \dfrac{.047}{.007} = 5.9 \geq 2.62$, so H_0 is rejected and this predictor is judged important.

d. $\hat{y} = 19.607 + .041(166) + .071(60) + .001(788) + .041(68) + .687(.95) = 99.514$, and the corresponding residual is $103 - 99.514 = 3.486$.

CHAPTER 14

Section 14.1

1.

    a. We reject H_0 if the calculated χ^2 value is greater than or equal to the tabled value of $\chi^2_{\alpha,k-1}$ from Table A.7. Since $12.25 \geq \chi^2_{.05,4} = 9.488$, we would reject H_0.

    b. Since 8.54 is not $\geq \chi^2_{.01,3} = 11.344$, we would fail to reject H_0.

    c. Since 4.36 is not $\geq \chi^2_{.10,2} = 4.605$, we would fail to reject H_0.

    d. Since 10.20 is not $\geq \chi^2_{.01,5} = 15.085$, we would fail to reject H_0.

3.   Using the number 1 for business, 2 for engineering, 3 for social science, and 4 for agriculture, let p_i = the true proportion of all clients from discipline i. If the Statistics department's expectations are correct, then the relevant null hypothesis is $H_0 : p_1 = .40, p_2 = .30, p_3 = .20, p_4 = .10$, versus H_a : the statistics department's expectations are not correct. With df $= k - 1 = 4 - 1 = 3$, we reject H_0 if $\chi^2 \geq \chi^2_{.05,3} = 7.815$. Using the proportions in H_0, the expected number of clients are :

Client's Discipline	Expected Number
Business	$(120)(.40) = 48$
Engineering	$(120)(.30) = 36$
Social Science	$(120)(.20) = 24$
Agriculture	$(120)(.10) = 12$

Since all the expected counts are at least 5, the chi-squared test can be used. The value of the test statistic is $\chi^2 = \sum_{i=1}^{k} \dfrac{(n_i - np_i)^2}{np_i} = \sum_{all\ cells} \dfrac{(obs - exp)^2}{exp}$

$= \left[\dfrac{(52-48)^2}{48} + \dfrac{(38-36)^2}{36} + \dfrac{(21-24)^2}{24} + \dfrac{(9-12)^2}{12} \right] = 1.57$, which is not ≥ 7.815, so we fail to reject H_0. (Alternatively, P-value $= P(\chi^2 \geq 1.57)$ which is $> .10$, and since the P-value is not $< .05$, we reject H_0). Thus we have no significant evidence to suggest that the statistics department's expectations are incorrect.

5. We will reject H_0 if the p-value $< .10$. The observed values, expected values, and corresponding χ^2 terms are :

Obs	4	15	23	25	38	21	32	14	10	8
Exp	6.67	13.33	20	26.67	33.33	33.33	26.67	20	13.33	6.67
χ^2	1.069	.209	.450	.105	.654	.163	1.065	1.800	.832	.265

$\chi^2 = 1.069 + ... + .265 = 6.612$. With df $= 10 - 1 = 9$, our χ^2 value of 6.612 is less than $\chi^2_{.10,9} = 14.684$, so the P-value $> .10$ and we cannot reject H_0. There is no significant evidence that the data is not consistent with the previously determined proportions.

7. We test $H_0 : p_1 = p_2 = p_3 = p_4 = .25$ vs. H_a : at least one proportion $\neq .25$, and df $= 3$. We will reject H_0 if the P-value $< .01$.

Cell	1	2	3	4
Observed	328	334	372	327
Expected	340.25	340.25	340.25	34.025
χ^2 term	.4410	.1148	2.9627	.5160

$\chi^2 = 4.0345$, and with 3 df, p-value $> .10$, so we fail to reject H_0. The data fails to indicate a seasonal relationship with incidence of violent crime.

9.

a. Denoting the 5 intervals by $[0, c_1), [c_1, c_2), ..., [c_4, \infty)$, we want c_1 for which

$.2 = P(0 \leq X \leq c_1) = \int_0^{c_1} e^{-x} dx = 1 - e^{-c_1}$, so $c_1 = -\ln(.8) = .2231$. Then

$.2 = P(c_1 \leq X \leq c_2) \Rightarrow .4 = P(0 \leq X_1 \leq c_2) = 1 - e^{-c_2}$, so $c_2 = -\ln(.6) = .5108$. Similarly, $c_3 = -\ln(.4) = .0163$ and $c_4 = -\ln(.2) = 1.6094$. the resulting intervals are $[0, .2231), [.2231, .5108), [.5108, .9163), [.9163, 1.6094)$, and $[1.6094, \infty)$.

b. Each expected cell count is $40(.2) = 8$, and the observed cell counts are 6, 8, 10, 7, and 9, so $\chi^2 = \left[\dfrac{(6-8)^2}{8} + ... + \dfrac{(9-8)^2}{8} \right] = 1.25$. Because 1.25 is not

$\geq \chi^2_{.10,4} = 7.779$, even at level .10 H_0 cannot be rejected; the data is consistent with the specified exponential distribution.

11.

a. The six intervals must be symmetric about 0, so denote the 4th, 5th and 6th intervals by $[0, a)$, $[a, b)$, $[b, \infty)$. The constant a must be such that $\Phi(a) = .6667(\frac{1}{2} + \frac{1}{6})$, which from Table A.3 gives $a \approx .43$. Similarly $\Phi(b) = .8333$ implies $b \approx .97$, so the six intervals are $(-\infty, -.97)$, $[-.97, -.43)$, $[-.43, 0)$, $[0, .43)$, $[.43, .97)$, and $[.97, \infty)$.

b. The six intervals are symmetric about the mean of .5. From **a**, the fourth interval should extend from the mean to .43 standard deviations above the mean, i.e., from .5 to $.5 + .43(.002)$, which gives $[.5, .50086)$. Thus the third interval is $[.5 - .00086, .5) = [.49914, .5)$. Similarly, the upper endpoint of the fifth interval is $.5 + .97(.002) = .50194$, and the lower endpoint of the second interval is $.5 - .00194 = .49806$. The resulting intervals are $(-\infty, .49806)$, $[.49806, .49914)$, $[.49914, .5)$, $[.5, .50086)$, $[.50086, .50194)$, and $[.50194, \infty)$.

c. Each expected count is $45(1/6) = 7.5$, and the observed counts are 13, 6, 6, 8, 7, and 5, so $\chi^2 = 5.53$. With 5 df, the P-value > .10, so we would fail to reject H_0 at any of the usual levels of significance. There is no statistically significant evidence to suggest that the bolt diameters are not normally distributed with $\mu = .5$ and $\sigma = .002$.

Section 14.2

13. According to the stated model, the three cell probabilities are $(1 - p)^2$, $2p(1 - p)$, and p^2, so we wish the value of p which maximizes $(1 - p)^{2n_1} [2p(1 - p)]^{n_2} p^{2n_3}$.

Proceeding as in example 14.6 gives $\hat{p} = \dfrac{n_2 + 2n_3}{2n} = \dfrac{234}{2776} = .0843$. The estimated expected cell counts are then $n(1 - \hat{p})^2 = 1163.85$, $n[2\hat{p}(1 - \hat{p})]^2 = 214.29$, $n\hat{p}^2 = 9.86$. This gives

$$\chi^2 = \left[\frac{(1212 - 1163.85)^2}{1163.85} + \frac{(118 - 214.29)^2}{214.29} + \frac{(58 - 9.86)^2}{9.86} \right] = 280.3 .$$ According to

(14.15), H_0 will be rejected if $\chi^2 \geq \chi^2_{\alpha,2}$, and since $\chi^2_{.01,2} = 9.210$, H_0 is soundly rejected; the stated model is strongly contradicted by the data.

15. The part of the likelihood involving θ is $\left[(1-\theta)^4\right]^{n_1} \cdot \left[\theta(1-\theta)^3\right]^{n_2} \cdot \left[\theta^2(1-\theta)^2\right]^{n_3} \cdot$

$\left[\theta^3(1-\theta)\right]^{n_4} \cdot \left[\theta^4\right]^{n_5} = \theta^{n_2+2n_3+3n_4+4n_5}(1-\theta)^{4n_1+3n_2+2n_3+n_4} = \theta^{233}(1-\theta)^{367}$, so the log-likelihood is $233\ln\theta + 367\ln(1-\theta)$. Differentiating and equating to 0 yields

$\hat{\theta} = \dfrac{233}{600} = .3883$, and $(1-\hat{\theta}) = .6117$ [note that the exponent on θ is simply the total

# of successes (defectives here) in the $n = 4(150) = 600$ trials]. Substituting this $\hat{\theta}$ into the formula for p_i yields estimated cell probabilities .1400, .3555, .3385, .1433, and .0227. Multiplication by 150 yields the estimated expected cell counts are 21.00, 53.33, 50.78, 21.50, and 3.41. the last estimated expected cell count is less than 5, so we combine the last two categories into a single one (≥ 3 defectives), yielding estimated counts 21.00, 53.33, 50.78, 24.91, observed counts 26, 51, 47, 26, and $\chi^2 = 1.62$. With df $= 4 - 1 - 1 = 2$, since $1.62 < \chi^2_{.10,2} = 4.605$, the P-value $> .10$, and we do not reject H_0. The data suggests that the stated binomial distribution is plausible.

17. $\hat{\lambda} = \dfrac{380}{120} = 3.167$, so $\hat{p} = e^{-3.167}\dfrac{(3.167)^x}{x!}$.

x	0	1	2	3	4	5	6	≥ 7
$\hat{p}$	.0421	.1334	.2113	.2230	.1766	.1119	.0590	.0427
$n\hat{p}$	5.05	16.00	25.36	26.76	21.19	13.43	7.08	5.12
obs	24	16	16	18	15	9	6	16

The resulting value of $\chi^2 = 103.98$, and when compared to $\chi^2_{.01,7} = 18.474$, it is obvious that the Poisson model fits very poorly.

19. With $A = 2n_1 + n_4 + n_5$, $B = 2n_2 + n_4 + n_6$, and $C = 2n_3 + n_5 + n_6$, the likelihood is proportional to $\theta_1^A \theta_2^B (1 - \theta_1 - \theta_2)^C$, where $A + B + C = 2n$. Taking the natural log and equating both $\dfrac{\partial}{\partial \theta_1}$ and $\dfrac{\partial}{\partial \theta_2}$ to zero gives $\dfrac{A}{\theta_1} = \dfrac{C}{1 - \theta_1 - \theta_2}$ and

    $\dfrac{B}{\theta_2} = \dfrac{C}{1 - \theta_1 - \theta_2}$, whence $\theta_2 = \dfrac{B\theta_1}{A}$. Substituting this into the first equation gives

    $\theta_1 = \dfrac{A}{A + B + C}$, and then $\theta_2 = \dfrac{B}{A + B + C}$. Thus $\hat{\theta}_1 = \dfrac{2n_1 + n_4 + n_5}{2n}$,

    $\hat{\theta}_2 = \dfrac{2n_2 + n_4 + n_6}{2n}$, and $\left(1 - \hat{\theta}_1 - \hat{\theta}_2\right) = \dfrac{2n_3 + n_5 + n_6}{2n}$. Substituting the observed

    n_i's yields $\hat{\theta}_1 = \dfrac{2(49) + 20 + 53}{400} = .4275$, $\hat{\theta}_2 = \dfrac{110}{400} = .2750$, and $\left(1 - \hat{\theta}_1 - \hat{\theta}_2\right) = .2975$,

    from which $\hat{p}_1 = (.4275)^2 = .183$, $\hat{p}_2 = .076$, $\hat{p}_3 = .089$, $\hat{p}_4 = 2(.4275)(.275) = .235$,
    $\hat{p}_5 = .254$, $\hat{p}_6 = .164$.

Category	1	2	3	4	5	6
np	36.6	15.2	17.8	47.0	50.8	32.8
observed	49	26	14	20	53	38

    This gives $\chi^2 = 29.1$. With $\chi^2_{.01, 6-1-2} = \chi^2_{.01, 3} = 11.344$, and
    $\chi^2_{.01, 6-1} = \chi^2_{.01, 5} = 15.085$, H_0 must be rejected since $29.1 \geq 15.085$.

21. The Ryan-Joiner test p-value is larger than .10, so we conclude that the null hypothesis of normality cannot be rejected. This data could reasonably have come from a normal population. This means that it would be legitimate to use a one-sample t test to test hypotheses about the true average ratio.

23. Minitab gives r = .967, though the hand calculated value may be slightly different because when there are ties among the $x_{(i)}$'s, Minitab uses the same y_i for each $x_{(i)}$ in a group of tied values. $C_{10} = .9707$, and $c_{.05} = 9639$, so $.05 < $ p-value $< .10$. At the 5% significance level, one would have to consider population normality plausible.

Section 14.3

25. Let P_{ij} = the proportion of white clover in area of type i which has a type j mark (i = 1, 2; j = 1, 2, 3, 4, 5). The hypothesis H_0: $p_{1j} = p_{2j}$ for j = 1, ..., 5 will be rejected at level .01 if $\chi^2 \geq \chi^2_{.01,(2-1)(5-1)} = \chi^2_{.01,4} = 13.277$.

$\hat{E}_{ij}$	1	2	3	4	5		
1	449.66	7.32	17.58	8.79	242.65	726	$\chi^2 = 23.18$
2	471.34	7.68	18.42	9.21	254.35	761	
	921	15	36	18	497	1487	

Since $23.18 \geq 13.277$, , H_0 is rejected.

27. With i = 1 identified with men and i = 2 identified with women, and j = 1, 2, 3 denoting the 3 categories L>R, L=R, L<R, we wish to test H_0: $p_{1j} = p_{2j}$ for j = 1, 2, 3 vs. H_a: p_{1j} not equal to p_{2j} for at least one j. The estimated cell counts for men are 17.95, 8.82, and 13.23 and for women are 39.05, 19.18, 28.77, resulting in $\chi^2 = 44.98$. With (2 – 1)(3 – 1) = 2 degrees of freedom, since

$44.98 > \chi^2_{.005,2} = 10.597$, p-value $< .005$, which strongly suggests that H_0 should be rejected.

29. H_0: $p_{1j} = ... = p_{6j}$ for j = 1, 2, 3 is the hypothesis of interest, where p_{ij} is the proportion of the jth sex combination resulting from the ith genotype. H_0 will be rejected at level .10 if $\chi^2 \geq \chi^2_{.10,10} = 15.987$.

$\hat{E}_{ij}$	1	2	3		χ^2	1	2	3	
1	35.8	83.1	35.1	154		.02	.12	.44	
2	39.5	91.8	38.7	170		.06	.66	1.01	
3	35.1	81.5	34.4	151		.13	.37	.34	
4	9.8	22.7	9.6	42		.32	.49	.26	
5	5.1	11.9	5.0	22		.00	.06	.19	
6	26.7	62.1	26.2	115		.40	.14	1.47	
	152	353	149	654					6.46

(carrying 2 decimal places in $\hat{E}_{ij}$ yields $\chi^2 = 6.49$). Since $6.46 < 15.987$, H_0 cannot be rejected at level .10.

31. With i denoting the ith type of car ($i = 1, 2, 3, 4$) and j the jth category of commuting distance, H_0: $p_{ij} = p_{i.}\, p_{.j}$ (type of car and commuting distance are independent) will be rejected at level .05 if $\chi^2 \geq \chi^2_{.05,6} = 12.592$.

$\hat{E}_{ij}$	1	2	3	
1	10.19	26.21	15.60	52
2	11.96	30.74	18.30	61
3	19.40	49.90	29.70	99
4	7.45	19.15	11.40	38
	49	126	75	250

$\chi^2 = 14.15 \geq 12.592$, so the independence hypothesis H_0 is rejected at level .05 (but not at level .025).

33. $\chi^2 = \Sigma\Sigma \dfrac{\left(N_{ij} - \hat{E}_{ij}\right)^2}{\hat{E}_{ij}} = \Sigma\Sigma \dfrac{N_{ij}^2 - 2\hat{E}_{ij}N_{ij} + \hat{E}_{ij}^2}{\hat{E}_{ij}} = \Sigma\Sigma \dfrac{N_{ij}^2}{\hat{E}_{ij}} - 2\Sigma\Sigma N_{ij} + \Sigma\Sigma\hat{E}_{ij}$, but

$\Sigma\Sigma\hat{E}_{ij} = \Sigma\Sigma N_{ij} = n$, so $\chi^2 = \Sigma\Sigma \dfrac{N_{ij}^2}{\hat{E}_{ij}} - n$. This formula is computationally efficient

because there is only one subtraction to be performed, which can be done as the last step in the calculation.

35. With p_{ij} denoting the common value of p_{ij1}, p_{ij2}, p_{ij3}, p_{ij4} (under H_0), $\hat{p}_{ij} = \dfrac{n_{ij.}}{n}$ and

$\hat{E}_{ijk} = \dfrac{n_k n_{ij.}}{n}$, where $n_{ij.} = \displaystyle\sum_{k=1}^{4} n_{ijk}$ and $n = \displaystyle\sum_{k=1}^{4} n_k$. With four different tables (one for

each region), there are $4(9-1) = 32$ freely determined cell counts. Under H_0, the nine parameters $p_{11}, \ldots, p_{33}$ must be estimated, but $\Sigma\Sigma p_{ij} = 1$, so only 8 independent parameters are estimated, giving χ^2 df $= 32 - 8 = 24$. Note: this is really a test of homogeneity for 4 strata, each with 3x3=9 categories. Hence, df $= (4-1)(9-1) = 24$.

Supplementary Exercises

37. There are 3 categories here – firstborn, middleborn, (2nd or 3rd born), and lastborn. With p_1, p_2, and p_3 denoting the category probabilities, we wish to test H_0: $p_1 = .25$, $p_2 = .50$ ($p_2 = P(2^{nd}$ or 3rd born) $= .25 + .25 = .50$), $p_3 = .25$. H_0 will be rejected at significance level .05 if $\chi^2 \geq \chi^2_{.05,2} = 5.992$. The expected counts are $(31)(.25) = 7.75$, $(31)(.50) = 15.5$, and 7.75, so $\chi^2 = \dfrac{(12 - 7.75)^2}{7.75} + \dfrac{(11 - 15.5)^2}{15.5} + \dfrac{(8 - 7.75)^2}{7.75} = 3.65$.

Because $3.65 < 5.992$, H_0 is not rejected. The hypothesis of equiprobable birth order appears quite plausible.

39. This is a test of homogeneity, with hypotheses H_0: proportions falling into these experience categories are the same for men and women, versus H_a: proportions falling into these experience categories are different for men and women. Df = 4, and we reject H_0 if $\chi^2 \geq \chi^2_{.01,4} = 13.277$.

	Years of Experience				
Gender	1 – 3	4 – 6	7 – 9	10 – 12	13 +
Male Observed	202	369	482	361	811
Expected	285.56	409.83	475.94	347.04	706.63
$\frac{(O-E)^2}{E}$	24.451	4.068	.077	.562	15.415
Female Observed	230	251	238	164	258
Expected	146.44	210.17	244.06	177.96	362.37
$\frac{(O-E)^2}{E}$	47.680	7.932	.151	1.095	30.061

$\chi^2 = \Sigma \dfrac{(O-E)^2}{E} = 131.492$. Reject H_0. The two variables do not appear to be independent. In particular, women have higher than expected counts in the beginning category (1 – 3 years) and lower than expected counts in the more experienced category (13+ years).

41. The null hypothesis H_0: $p_{ij} = p_{i.} \, p_{.j}$ states that level of parental use and level of student use are independent in the population of interest. The test is based on $(3-1)(3-1) = 4$ df.

Estimated Expected

119.3	57.6	58.1	235
82.8	33.9	40.3	163
23.9	11.5	11.6	47
226	109	110	445

The calculated value of $\chi^2 = 22.4$. Since $22.4 > \chi^2_{.005,4} = 14.860$, p-value $< .005$, so H_0 should be rejected at any significance level greater than .005. Parental and student use level do not appear to be independent.

43. This is a test of homogeneity: H_0: $p_{1j} = p_{2j} = p_{3j}$ for $j = 1, 2, 3, 4, 5$. The given SPSS output reports the calculated $\chi^2 = 70.64156$ and accompanying p-value of .0000. We reject H_0 at any significance level. The data strongly supports that there are differences in perception of odors among the three areas.

45. $(n_1 - np_{10})^2 = (np_{10} - n_1)^2 = (n - n_1 - n(1 - p_{10}))^2 = (n_2 - np_{20})^2$. Therefore

$$\chi^2 = \frac{(n_1 - np_{10})^2}{np_{10}} + \frac{(n_2 - np_{20})^2}{np_{20}} = \frac{(n_1 - np_{10})^2}{n_2}\left(\frac{n}{p_{10}} + \frac{n}{p_{20}}\right)$$

$$= \left(\frac{n_1}{n} - p_{10}\right)^2 \cdot \left(\frac{n}{p_{10}\,p_{20}}\right) = \frac{(\hat{p}_1 - p_{10})^2}{p_{10}p_{20}/n} = z^2.$$

47.

a. Our hypotheses are H_0: no difference in proportion of concussions among the three groups, versus H_a: there is a difference in proportion of concussions among the three groups.

Observed	Concussion	No Concussion	Total
Soccer	45	46	91
Non Soccer	28	68	96
Control	8	45	53
Total	81	159	240

Expected	Concussion	No Concussion	Total
Soccer	30.7125	60.2875	91
Non Soccer	32.4	63.6	96
Control	17.8875	37.1125	53
Total	81	159	240

$$\chi^2 = \frac{(45-30.7125)^2}{30.7125} + \frac{(46-60.2875)^2}{60.2875} + \frac{(28-32.4)^2}{32.4} + \frac{(68-63.6)^2}{63.6}$$

$$+ \frac{(8-17.8875)^2}{17.8875} + \frac{(45-37.1125)^2}{37.1125} = 19.1842 .$$ The df for this test is $(I-1)(J-1) = 2$, so we reject H_0 if $\chi^2 > \chi^2_{.05,2} = 5.99$. $19.1842 > 5.99$, so we reject H_0.

There is a difference in the proportion of concussions based on whether a person plays soccer.

b. We are testing the hypothesis H_0: $\rho = 0$ vs H_a: $\rho \neq 0$. The test statistic is

$$t = \frac{r\sqrt{n-2}}{\sqrt{1-r^2}} = \frac{-.22\sqrt{89}}{\sqrt{1-.22^2}} = -2.13 .$$ At significance level $\alpha = .01$, we would fail to reject and conclude that there is no evidence of non-zero correlation in the population. If we were willing to accept a higher significance level, our decision could change. At best, there is evidence of only weak correlation.

c. We will test to see if the average score on a controlled word association test is the same for soccer and non-soccer athletes. $H_0: \mu_1 = \mu_2$ vs $H_a: \mu_1 \neq \mu_2$. We'll use

test statistic $t = \dfrac{(\bar{x}_1 - \bar{x}_2)}{\sqrt{\dfrac{s_1^2}{m} + \dfrac{s_2^2}{n}}}$. With $\dfrac{s_1^2}{m} = 3.206$ and $\dfrac{s_2^2}{n} = 1.854$,

$t = \dfrac{(37.50 - 39.63)}{\sqrt{3.206 + 1.854}} = -.95$. Estimated df $= \dfrac{(3.206 + 1.854)^2}{\dfrac{3.206^2}{25} + \dfrac{1.854^2}{55}} \approx 56$. The p-value

is $> .10$, so we do not reject H_0 and conclude that there is no significant difference in the average score on the test for the two groups of athletes.

d. Our hypotheses for ANOVA are H_0: all means are equal vs H_a: not all means are equal. The test statistic is $f = \dfrac{MSTr}{MSE}$.

$SSTr = 91(.30 - .35)^2 + 96(.49 - .35)^2 + 53(.19 - .35)^2 = 3.4659$

$MSTr = \dfrac{3.4659}{2} = 1.73295$ $SSE = 90(.67)^2 + 95(.87)^2 + 52(.48)^2 = 124.2873$

and $MSE = \dfrac{124.2873}{237} = .5244$. Now, $f = \dfrac{1.73295}{.5244} = 3.30$. Using df $= (2,200)$

from table A.9, the p value is between .01 and .05. At significance level .05, we reject the null hypothesis. There is sufficient evidence to conclude that there is a difference in the average number of prior non-soccer concussions between the three groups.

49.

a. No: A normal distribution with mean 1.07 and standard deviation 1.83 would include negative values. Clearly, the number of concussions a person has had cannot be negative!

b. Let μ = the true average number of concussions among soccer players. Since $n = 91$, we can construct a t confidence interval for μ. For example, a 99% CI for μ is

$\bar{x} \pm t_{.005,90} \dfrac{s}{\sqrt{n}} = 1.07 \pm 2.63 \dfrac{1.83}{\sqrt{91}} \approx (0.57, 1.57)$. We are 95% confident that the

true average number of concussions among all soccer players is between 0.57 and 1.57.

c. We wish to perform a chi-squared test of homogeneity. With p_i = the true probability of concussion ($i = 1$ for soccer players, 2 for nonsoccer athletes, 3 for controls), we wish to test the null hypothesis H_0: $p_1 = p_2 = p_3$. From software, the test statistic is $\chi^2 = 19.184$, and the associated P-value at df = $(3-1)(2-1) = 2$ is 0.000. Hence, we strongly reject the null hypothesis. The true proportions of individuals in these three groups who have had a concussion are not identical.

d. Let μ_i = the true average Weschler digit span test score for these groups ($i = 1$, 2, 3). We wish to test H_0: $\mu_1 = \mu_2 = \mu_3$. This requires analysis of variance. Using software, the test statistic is $f = 1.35$, and the associated P-value at df = $(2,231)$ is around 0.874. Hence, we clearly fail to reject the null hypothesis. The data does not suggest any significant difference in the true average digit span test scores of these three groups.

CHAPTER 15

Section 15.1

1. We test $H_0 : \mu = 100$ vs. $H_a : \mu \neq 100$. The test statistic is s_+ = sum of the ranks associated with the positive values of $(x_i - 100)$, and we reject H_0 at significance level .05 if $s_+ \geq 64$. (from Table A.13, n = 12, with $\alpha/2 = .026$, which is close to the desired value of .025), or if $s_+ \leq \dfrac{12(13)}{2} - 64 = 78 - 64 = 14$.

x_i	$(x_i - 100)$	ranks
105.6	5.6	7*
90.9	−9.1	12
91.2	−8.8	11
96.9	−3.1	3
96.5	−3.5	5
91.3	−8.7	10
100.1	0.1	1*
105	5	6*
99.6	−0.4	2
107.7	7.7	9*
103.3	3.3	4*
92.4	−7.6	8

   $S_+ = 27$, and since 27 is neither ≥ 64 nor ≤ 14, we do not reject H_0. There is not enough evidence to suggest that the mean is something other than 100.

3. We test $H_0 : \mu = 7.39$ v. $H_a : \mu \neq 7.39$, so a two tailed test is appropriate. With n = 14 and $\alpha/2 = .025$, Table A.13 indicates that H_0 should be rejected if either $s_+ \geq 84$ or $s_+ \leq 21$. The $(x_i - 7.39)$'s are −.37, −.04, −.05, −.22, −.11, .38, −.30, −.17, .06, −.44, .01, −.29, −.07, and −.25, from which the ranks of the three positive differences are 1, 4, and 13. Since $s_+ = 18 \leq 21$, H_0 is rejected at level .05.

Chapter 15: Distribution–Free Procedures

5. The data is paired, and we wish to test $H_0 : \mu_D = 0$ vs. $H_a : \mu_D \neq 0$. With n = 12 and α = .05, H_0 should be rejected if either $s_+ \geq 64$ or if $s_+ \leq 14$.

d_i	−.3	2.8	3.9	.6	1.2	−1.1	2.9	1.8	.5	2.3	.9	2.5
rank	1	10*	12*	3*	6*	5	11*	7*	2*	8*	4*	9*

$s_+ = 72 \geq 64$, so H_0 is rejected at level .05. In fact for α = .01, the critical value is c = 71, so even at level .01 H_0 would be rejected.

7. $H_0 : \mu_D = .20$ vs. $H_a : \mu_D > .20$, where $\mu_D = \mu_{outdoor} - \mu_{indoor}$. $\alpha = .05$, and because n = 33, we can use the large sample test. The test statistic is

$$Z = \frac{s_+ - \frac{n(n+1)}{4}}{\sqrt{\frac{n(n+1)(2n+1)}{24}}}, \text{ and we reject } H_0 \text{ if } z \geq 1.96.$$

d_i	$d_i - .2$	rank	d_i	$d_i - .2$	rank	d_i	$d_i - .2$	rank
0.22	0.02	2	0.15	−0.05	5.5	0.63	0.43	23
0.01	−0.19	17	1.37	1.17	32	0.23	0.03	4
0.38	0.18	16	0.48	0.28	21	0.96	0.76	31
0.42	0.22	19	0.11	−0.09	8	0.2	0	1
0.85	0.65	29	0.03	−0.17	15	−0.02	−0.22	18
0.23	0.03	3	0.83	0.63	28	0.03	−0.17	14
0.36	0.16	13	1.39	1.19	33	0.87	0.67	30
0.7	0.5	26	0.68	0.48	25	0.3	0.1	9.5
0.71	0.51	27	0.3	0.1	9.5	0.31	0.11	11
0.13	−0.07	7	−0.11	−0.31	22	0.45	0.25	20
0.15	−0.05	5.5	0.31	0.11	12	−0.26	−0.46	24

$s_+ = 424$, so $z = \dfrac{424 - 280.5}{\sqrt{3132.25}} = \dfrac{143.5}{55.9665} = 2.56$. Since $2.56 \geq 1.96$, we reject H_0 at significance level .05.

9.

r_1	1	1	1	1	1	1	2	2	2	2	2	2
r_2	2	2	3	3	4	4	1	1	3	3	4	4
r_3	3	4	2	4	2	3	3	4	1	4	1	3
r_4	4	3	4	2	3	2	4	3	4	1	3	1
D	0	2	2	6	6	8	2	4	6	12	10	14

r_1	3	3	3	3	3	3	4	4	4	4	4	4
r_2	1	1	2	2	4	4	1	1	2	2	3	3
r_3	2	4	1	4	1	2	2	3	1	3	1	2
r_4	4	2	4	1	2	1	3	2	3	1	2	1
D	6	10	8	14	16	18	12	14	14	18	18	20

When H_0 is true, each of the above 24 rank sequences is equally likely, which yields the distribution of D when H_0 is true as described in the answer section; e.g., $P(D = 2)$ = P(1243 or 1324 or 2134) = 3/24. Then $c = 0$ yields $\alpha = 1/24 = .042$ while $c = 2$ implies $\alpha = 4/24 = .167$.

Section 15.2

11. With X identified with pine (corresponding to the smaller sample size) and Y with oak, we wish to test $H_0 : \mu_1 - \mu_2 = 0$ vs. $H_a : \mu_1 - \mu_2 \neq 0$. From Table A.14 with m = 6 and n = 8, H_0 is rejected in favor of H_a at level .05 if either $w \geq 61$ or if $w \leq 90$ − 61 = 29 (the actual α is 2(.021) = .042). The X ranks are 3 (for .73), 4 (for .98), 5 (for 1.20), 7 (for 1.33), 8 (for 1.40), and 10 (for 1.52), so w = 37. Since 37 is neither ≥ 61 nor ≤ 29, H_0 cannot be rejected.

13. Here m = n = 10 > 8, so we use the large-sample test statistic from pp. 680–681. $H_0 : \mu_1 - \mu_2 = 0$ will be rejected at level .01 in favor of $H_a : \mu_1 - \mu_2 \neq 0$ if either $z \geq 2.58$ or $z \leq -2.58$. Identifying X with orange juice, the X ranks are 7, 8, 9, 10, 11, 16, 17, 18, 19, and 20, so w = 135. With $\dfrac{m(m+n+1)}{2} = 105$ and

$\sqrt{\dfrac{mn(m+n+1)}{12}} = \sqrt{175} = 13.22$, $z = \dfrac{135-105}{13.22} = 2.27$. Because 2.27 is neither ≥ 2.58 nor ≤ -2.58, H_0 is not rejected. P-value $\approx 2(1 - \Phi(2.27)) = .0232$.

15. Let μ_1 and μ_2 denote true average cotanine levels in unexposed and exposed infants, respectively. The hypotheses of interest are $H_0 : \mu_1 - \mu_2 = -25$ vs. $H_a : \mu_1 - \mu_2 < -25$. With m = 7, n = 8, H_0 will be rejected at level .05 if $w \leq 7(7+8+1) - 71 = 41$. Before ranking, –25 is subtracted from each x_i (i.e. 25 is added to each), giving 33, 36, 37, 39, 45, 68, and 136. The corresponding ranks in the combined set of 15 observations are 1, 3, 4, 5, 6, 8, and 12, from which $w = 1 + 3 + \ldots + 12 = 39$. Because $39 \leq 41$, H_0 is rejected. The true average level for exposed infants appears to exceed that for unexposed infants by more than 25 (note that H_0 would not be rejected using level .01).

Section 15.3

17. $n = 8$, so from Table A.15, a 95% C.I. (actually 94.5%) has the form $(\bar{x}_{(36-32+1)}, \bar{x}_{(32)}) = (\bar{x}_{(5)}, \bar{x}_{(32)})$. It is easily verified that the 5 smallest pairwise averages are $\dfrac{5.0 + 5.0}{2} = 5.00$, $\dfrac{5.0 + 11.8}{2} = 8.40$, $\dfrac{5.0 + 12.2}{2} = 8.60$, $\dfrac{5.0 + 17.0}{2} = 11.00$, and $\dfrac{5.0 + 17.3}{2} = 11.15$ (the smallest average not involving 5.0 is $\bar{x}_{(6)} = \dfrac{11.8 + 11.8}{2} = 11.8$), and the 5 largest averages are 30.6, 26.0, 24.7, 23.95, and 23.80, so the confidence interval is (11.15, 23.80).

19. With $n = 8$, Table A.15 gives c = 32, and so a 94.5% confidence interval for μ_D is $(\bar{x}_{(5)}, \bar{x}_{(32)})$. The five smallest pairwise averages are –.71, –.68, –.65, –.615, –.585; the five largest pairwise averages are .2, .195, .19, .03, .025. The desired confidence interval is thus (–.585, .025).

21. $m = n = 5$ and from Table A.16, c = 21 and the 90% (actually 90.5%) interval is $(d_{ij(5)}, d_{ij(21)})$. The five smallest $x_i - y_j$ differences are –18, –2, 3, 4, 16 while the five largest differences are 136, 123, 120, 107, 87, so the desired interval is (16, 87).

Section 15.4

23. Below we record in parentheses beside each observation the rank of that observation in the combined sample.

1:	5.8(3)	6.1(5)	6.4(6)	6.5(7)	7.7(10)	$r_{1.} = 31$
2:	7.1(9)	8.8(12)	9.9(14)	10.5(16)	11.2(17)	$r_{2.} = 68$
3:	5.1(1)	5.7(2)	5.9(4)	6.6(8)	8.2(11)	$r_{3.} = 26$
4:	9.5(13)	1.0.3(15)	11.7(18)	12.1(19)	12.4(20)	$r_{4.} = 85$

H_0 will be rejected at level .10 if $k \geq \chi^2_{.10,3} = 6.251$. The computed value of k is

$$k = \frac{12}{20(21)}\left[\frac{31^2 + 68^2 + 26^2 + 85^2}{5}\right] - 3(21) = 14.06. \text{ Since } 14.06 \geq 6.251, \text{ reject } H_0.$$

25. $H_0 : \mu_1 = \mu_2 = \mu_3$ will be rejected at level .05 if $k \geq \chi^2_{.05,2} = 5.992$. The ranks are 1, 3, 4, 5, 6, 7, 8, 9, 12, 14 for the first sample; 11, 13, 15, 16, 17, 18 for the second; 2, 10, 19, 20, 21, 22 for the third; so the rank totals are 69, 90, and 94.

$$k = \frac{12}{22(23)}\left[\frac{69^2}{10} + \frac{90^2}{6} + \frac{94^2}{5}\right] - 3(23) = 9.23. \text{ Since } 9.23 \geq 5.992, \text{ we reject } H_0.$$

27.

	1	2	3	4	5	6	7	8	9	10	r_i	r_i^2
I	1	2	3	3	2	1	1	3	1	2	19	361
H	2	1	1	2	1	2	2	1	2	3	17	289
C	3	3	2	1	3	3	3	2	3	1	24	576
												1226

The computed value of F_r is $\dfrac{12}{10(3)(4)}(1226) - 3(10)(4) = 2.60$, which is not

$\geq \chi^2_{.05,2} = 5.992$, so don't reject H_0.

Supplementary Exercises

29. Friedman's test is appropriate here. At level .05, H_0 will be rejected if
$f_r \geq \chi^2_{.05,3} = 7.815$. It is easily verified that $r_{1.} = 28$, $r_{2.} = 29$, $r_{3.} = 16$, $r_{4.} = 17$,
from which the defining formula gives $f_r = 9.62$ and the computing formula gives
$f_r = 9.67$. Because $f_r \geq 7.815$, $H_0 : \alpha_1 = \alpha_2 = \alpha_3 = \alpha_4 = 0$ is rejected, and we
conclude that there are effects due to different years.

31. From Table A.16, $m = n = 5$ implies that $c = 22$ for a confidence level of 95%, so
$mn - c + 1 = 25 - 22 = 1 = 4$. Thus the confidence interval extends from the 4th
smallest difference to the 4th largest difference. The 4 smallest differences are -7.1,
-6.5, -6.1, -5.9, and the 4 largest are -3.8, -3.7, -3.4, -3.2, so the CI is $(-5.9, -3.8)$.

33.

a. With "success" as defined, then Y is a binomial with n = 20. To determine the
binomial proportion "p" we realize that since 25 is the hypothesized median, 50%
of the distribution should be above 25, thus p = .50. From the binomial tables
(Table A.1) with n = 20 and p = .50, we see that
$\alpha = P(Y \geq 15) = 1 - P(Y \leq 14) = 1 - .979 = .021$.

b. From the same binomial table as in **a**, we find that
$P(Y \geq 14) = 1 - P(Y \leq 13) = 1 - .942 = .058$ (as close as we can get to .05), so c
= 14. For this data, we would reject H_0 at level .058 if $Y \geq 14$. Y = the number of
observations in the sample that exceed 25 = 12, and since 12 is not ≥ 14, we fail
to reject H_0.

35.

Sample:	y	x	y	y	x	x	x	y	y
Observations:	3.7	4.0	4.1	4.3	4.4	4.8	4.9	5.1	5.6
Rank:	1	3	5	7	9	8	6	4	2

The value of W' for this data is $w' = 3 + 6 + 8 + 9 = 26$. At level .05, the critical
value for the upper–tailed test is (Table A.14, $m = 4$, $n = 5$) $c = 27$ ($\alpha = .056$). Since
26 is not ≥ 27, H_0 cannot be rejected at level .05.

CHAPTER 16

Section 16.1

1. All ten values of the quality statistic are between the two control limits, so no out–of–control signal is generated.

3. P(10 successive points inside the limits) = P(1st inside) x P(2nd inside) x…x P(10th inside) = $(.998)^{10}$ = .9802. P(25 successive points inside the limits) = $(.998)^{25}$ = .9512. $(.998)^{52}$ = .9011, but $(.998)^{53}$ = .8993, so for 53 successive points the probability that at least one will fall outside the control limits when the process is in control is $1 - .8993 = .1007 > .10$.

Section 16.2

5.
a. P(point falls outside the limits when $\mu = \mu_0 + .5\sigma$)

$$= 1 - P\left(\mu_0 - \frac{3\sigma}{\sqrt{n}} < \overline{X} < \mu_0 + \frac{3\sigma}{\sqrt{n}} \text{ when } \mu = \mu_0 + .5\sigma\right)$$

$$= 1 - P\left(-3 - .5\sqrt{n} < Z < 3 - .5\sqrt{n}\right)$$

$$= 1 - P\left(-4.12 < Z < 1.882\right) = 1 - .9699 = .0301 .$$

b. $1 - P\left(\mu_0 - \frac{3\sigma}{\sqrt{n}} < \overline{X} < \mu_0 + \frac{3\sigma}{\sqrt{n}} \text{ when } \mu = \mu_0 - \sigma\right) = 1 - P\left(-3 + \sqrt{n} < Z < 3 + \sqrt{n}\right)$

$$= 1 - P\left(-.76 < Z < 5.24\right) = .2236$$

c. $1 - P\left(-3 - 2\sqrt{n} < Z < 3 - 2\sqrt{n}\right) = 1 - P\left(-7.47 < Z < -1.47\right) = .9292$

7. $\overline{\overline{x}} = 12.95$ and $\overline{s} = .526$, so with $a_5 = .940$, the control limits are

$$12.95 \pm 3\frac{.526}{.940\sqrt{5}} = 12.95 \pm .75 = 12.20, 13.70 .$$ Again, every point $(\overline{x})$ is between

these limits, so there is no evidence of an out–of–control process.

Chapter 16: Quality Control Methods

9. $\bar{\bar{x}} = \dfrac{2317.07}{24} = 96.54$, $\bar{s} = 1.264$, and $a_6 = .952$, giving the control limits

$96.54 \pm 3\dfrac{1.264}{.952\sqrt{6}} = 96.54 \pm 1.63 = 94.91, 98.17$. The value of $\bar{x}$ on the 22nd day lies

above the UCL, so the process appears to be out of control at that time.

11.

   a. $P\left(\mu_0 - \dfrac{2.81\sigma}{\sqrt{n}} < \bar{X} < \mu_0 + \dfrac{2.81\sigma}{\sqrt{n}} \text{ when } \mu = \mu_0\right) = P(-2.81 < Z < 2.81) = .995$,

   so the probability that a point falls outside the limits is .005 and

   $ARL = \dfrac{1}{.005} = 200$.

   b. $P = P(\text{a point is outside the limits})$

   $= 1 - P\left(\mu_0 - \dfrac{2.81\sigma}{\sqrt{n}} < \bar{X} < \mu_0 + \dfrac{2.81\sigma}{\sqrt{n}} \text{ when } \mu = \mu_0 + \sigma\right)$

   $= 1 - P\left(-2.81 - \sqrt{n} < Z < 2.81 - \sqrt{n}\right) = 1 - P(-4.81 < Z < .81) = 1 - .791 = .209$.

   Thus $ARL = \dfrac{1}{.209} = 4.78$

   c. $1 - .9974 = .0026$ so $ARL = \dfrac{1}{.0026} = 385$ for an in–control process, and when

   $\mu = \mu_0 + \sigma$, the probability of an out–of–control point is $1 - P(-3 - 2 < Z < 1)$

   $= 1 - P(Z < 1) = .1587$, so $ARL = \dfrac{1}{.1587} = 6.30$.

13. $\bar{\bar{x}} = 12.95$, IQR $= .4273$, $k_5 = .990$. The control limits are

$12.95 \pm 3\dfrac{.4273}{.990\sqrt{5}} = 12.37, 13.53$.

276

Section 16.3

15.

    a. $\bar{r} = \dfrac{85.2}{30} = 2.84, b_4 = 2.058$, and $c_4 = .880$. Since $n = 4$, LCL $= 0$ and UCL

    $= 2.84 + \dfrac{3(.880)(2.84)}{2.058} = 2.84 + 3.64 = 6.48$.

    b. $\bar{r} = 3.54, b_8 = 2.844$, and $c_8 = .820$, and the control limits are

    $= 3.54 \pm \dfrac{3(.820)(3.54)}{2.844} = 3.54 \pm 3.06 = .48, 6.60$.

17. $\bar{s} = 1.2642$, $a_6 = .952$, and the control limits are

    $1.2642 \pm \dfrac{3(1.2642)\sqrt{1 - (.952)^2}}{.952} = 1.2642 \pm 1.2194 = .045, 2.484$. The smallest s_i is s_{20}

    $= .75$, and the largest is $s_{12} = 1.65$, so every value is between .045 and 2.434. The
process appears to be in control with respect to variability.

Section 16.4

19. $\bar{p} = \Sigma \dfrac{\hat{p}_i}{k}$ where $\Sigma \hat{p}_i = \dfrac{x_1}{n} + \ldots + \dfrac{x_k}{n} = \dfrac{x_1 + \ldots + x_k}{n} = \dfrac{578}{100} = 5.78$. Thus

    $\bar{p} = \dfrac{5.78}{25} = .231$.

    a. The control limits are $.231 \pm 3\sqrt{\dfrac{(.231)(.769)}{100}} = .231 \pm .126 = .105, .357$.

    b. $\dfrac{13}{100} = .130$, which is between the limits, but $\dfrac{39}{100} = .390$, which exceeds the
upper control limit and therefore generates an out–of–control signal.

21.   LCL > 0 when $\bar{p} > 3\sqrt{\dfrac{\bar{p}(1 - \bar{p})}{n}}$, i.e. (after squaring both sides) $50\bar{p}^2 > 3\bar{p}(1 - \bar{p})$, i.e.

    $50\bar{p} > 3(1 - \bar{p})$, i.e. $53\bar{p} > 3 \Rightarrow \bar{p} = \dfrac{3}{53} = .0566$.

23. $\Sigma x_i = 102$, $\bar{x} = 4.08$, and $\bar{x} \pm 3\sqrt{\bar{x}} = 4.08 \pm 6.06 \approx (-2.0, 10.1)$. Thus LCL = 0 and UCL = 10.1. Because no x_i exceeds 10.1, the process is judged to be in control.

25. With $u_i = \dfrac{x_i}{g_i}$, the u_i's are 3.75, 3.33, 3.75, 2.50, 5.00, 5.00, 12.50, 12.00, 6.67, 3.33, 1.67, 3.75, 6.25, 4.00, 6.00, 12.00, 3.75, 5.00, 8.33, and 1.67 for $i = 1, \ldots, 20$, giving $\bar{u} = 5.5125$. For $g_i = .6$, $\bar{u} \pm 3\sqrt{\dfrac{\bar{u}}{g_i}} = 5.5125 \pm 9.0933$, LCL = 0, UCL = 14.6.

For $g_i = .8$, $\bar{u} \pm 3\sqrt{\dfrac{\bar{u}}{g_i}} = 5.5125 \pm 7.857$, LCL = 0, UCL = 13.4. For $g_i = 1.0$,

$\bar{u} \pm 3\sqrt{\dfrac{\bar{u}}{g_i}} = 5.5125 \pm 7.0436$, LCL = 0, UCL = 12.6. Several u_i's are close to the corresponding UCL's but none exceed them, so the process is judged to be in control.

Section 16.5

27. $\mu_0 = 16$, $k = \dfrac{\Delta}{2} = 0.05$, $h = .20$, $d_i = \max\left(0, d_{i-1} + (\bar{x}_i - 16.05)\right)$,

$e_i = \max\left(0, e_{i-1} + (\bar{x}_i - 15.95)\right)$.

i	$\bar{x}_i - 16.05$	d_i	$\bar{x}_i - 15.95$	e_i
1	−0.058	0	0.024	0
2	0.001	0.001	0.101	0
3	0.016	0.017	0.116	0
4	−0.138	0	−0.038	0.038
5	−0.020	0	0.080	0
6	0.010	0.010	0.110	0
7	−0.068	0	0.032	0
8	−0.151	0	−0.054	0.054
9	−0.012	0	0.088	0
10	0.024	0.024	0.124	0
11	−0.021	0.003	0.079	0
12	−0.115	0	−0.015	0.015
13	−0.018	0	0.082	0
14	−0.090	0	0.010	0
15	0.005	0.005	0.105	0

For no time r is it the case that $d_r > .20$ or that $e_r > .20$, so no out-of-control signals are generated.

29. Connecting 600 on the in-control ARL scale to 4 on the out-of-control scale and extending to the k' scale gives $k' = .87$. Thus $k' = \dfrac{\Delta/2}{\sigma/\sqrt{n}} = \dfrac{.002}{.005/\sqrt{n}}$ from which

$\sqrt{n} = 2.175 \Rightarrow n = 4.73 = s$. Then connecting .87 on the k' scale to 600 on the out-of-control ARL scale and extending to h' gives $h' = 2.8$, so

$$h = \left(\frac{\sigma}{\sqrt{n}}\right)(2.8) = \left(\frac{.005}{\sqrt{5}}\right)(2.8) = .00626 .$$

Section 16.6

31. For the binomial calculation, n = 50 and we wish

$$P(X \le 2) = \binom{50}{0} p^0 (1-p)^{50} + \binom{50}{1} p^1 (1-p)^{49} + \binom{50}{2} p^2 (1-p)^{48}$$

$$= (1-p)^{50} + 50p(1-p)^{49} + 1225p^2(1-p)^{48} \text{ when } p = .01, .02, ..., .10. \text{ For the}$$

hypergeometric calculation

$$P(X \le 2) = \frac{\binom{M}{0}\binom{500-M}{50}}{\binom{500}{50}} + \frac{\binom{M}{1}\binom{500-M}{49}}{\binom{500}{50}} + \frac{\binom{M}{2}\binom{500-M}{48}}{\binom{500}{50}}, \text{ to be calculated}$$

for M = 5, 10, 15, ..., 50. The resulting probabilities appear in the answer section in the text.

33.
$$P(X \le 2) = \binom{100}{0} p^0 (1-p)^{100} + \binom{100}{1} p^1 (1-p)^{99} + \binom{100}{2} p^2 (1-p)^{98}$$

p	.01	.02	.03	.04	.05	.06	.07	.08	.09	.10
$P(X \le 2)$	.9206	.6767	.4198	.2321	.1183	.0566	.0258	.0113	.0048	.0019

For values of p quite close to 0, the probability of lot acceptance using this plan is larger than that for the previous plan, whereas for larger p this plan is less likely to result in an "accept the lot" decision (the dividing point between "close to zero" and "larger p" is someplace between .01 and .02). In this sense, the current plan is better.

35. P(accept the lot) = P(X_1 = 0, 1) + P(X_1 = 2, X_2 = 0, 1, 2, 3) + P(X_1 = 3, X_2 = 0, 1, 2) = P(X_1 = 0, 1) + P(X_1 = 2)P(X_2 = 0, 1, 2, 3) + P(X_1 = 3)P(X_2 = 0, 1, 2).

$p = .01$: P(accept the lot) = .9106 + (.0756)(.9984) + (.0122)(.9862) = .9981

$p = .05$: P(accept the lot) = .2794 + (.2611)(.7604) + (.2199)(.5405) = .5968

$p = .10$: P(accept the lot) = .0338 + (.0779)(.2503) + (.1386)(.1117) = .0688

37.

a. $AOQ = pP(A) = p[(1-p)^{50} + 50p(1-p)^{49} + 1225p^2(1-p)^{48}]$

p	.01	.02	.03	.04	.05	.06	.07	.08	.09	.10
AOQ	.010	.018	.024	.027	.027	.025	.022	.018	.014	.011

b. $p = .0447$, $AOQL = .0447P(A) = .0274$

c. $ATI = 50P(A) + 2000(1 - P(A))$

p	.01	.02	.03	.04	.05	.06	.07	.08	.09	.10
ATI	77.3	202.1	418.6	679.9	945.1	1188.8	1393.6	1559.3	1686.1	1781.6

Supplementary Exercises

39. $n = 6$, $k = 26$, $\Sigma \bar{x}_i = 10,980$, $\bar{\bar{x}} = 422.31$, $\Sigma s_i = 402$, $\bar{s} = 15.4615$, $\Sigma r_i = 1074$,
$\bar{r} = 41.3077$

S chart: $15.4615 \pm \dfrac{3(15.4615)\sqrt{1-(.952)^2}}{.952} = 15.4615 \pm 14.9141 \approx .55, 30.37$

R chart: $41.31 \pm \dfrac{3(.848)(41.31)}{2.536} = 41.31 \pm 41.44$, so LCL = 0, UCL = 82.75

$\bar{X}$ chart based on $\bar{s}$: $422.31 \pm \dfrac{3(15.4615)}{.952\sqrt{6}} = 402.42, 442.20$

$\bar{X}$ chart based on $\bar{r}$: $422.31 \pm \dfrac{3(41.3077)}{2.536\sqrt{6}} = 402.36, 442.26$

41.

i	$\bar{x}_i$	s_i	r_i
1	50.83	1.172	2.2
2	50.10	.854	1.7
3	50.30	1.136	2.1
4	50.23	1.097	2.1
5	50.33	.666	1.3
6	51.20	.854	1.7
7	50.17	.416	.8
8	50.70	.964	1.8
9	49.93	1.159	2.1
10	49.97	.473	.9
11	50.13	.698	.9
12	49.33	.833	1.6
13	50.23	.839	1.5
14	50.33	.404	.8
15	49.30	.265	.5
16	49.90	.854	1.7
17	50.40	.781	1.4
18	49.37	.902	1.8
19	49.87	.643	1.2
20	50.00	.794	1.5
21	50.80	2.931	5.6
22	50.43	.971	1.9

$\Sigma s_i = 19.706$, $\bar{s} = .8957$, $\Sigma \bar{x}_i = 1103.85$, $\bar{\bar{x}} = 50.175$, $a_3 = .886$, from which an s chart has LCL $= 0$ and UCL $= .8957 + \dfrac{3(.8957)\sqrt{1-(.886)^2}}{.886} = 2.3020$, and

$s_{21} = 2.931 > UCL$. Since an assignable cause is assumed to have been identified we eliminate the 21st group. Then $\Sigma s_i = 16.775$, $\bar{s} = .7998$, $\bar{\bar{x}} = 50.145$. The resulting UCL for an s chart is 2.0529, and $s_i < 2.0529$ for every remaining i. The $\bar{x}$ chart based on $\bar{s}$ has limits $50.145 \pm \dfrac{3(.7988)}{.886\sqrt{3}} = 48.58, 51.71$. All $\bar{x}_i$ values are between these limits.

43. $\Sigma n_i = 4(16) + (3)(4) = 76$, $\Sigma n_i \bar{x}_i = 32{,}729.4$, $\bar{\bar{x}} = 430.65$,

$$s^2 = \frac{\Sigma(n_i - 1)s_i^2}{\Sigma(n_i - 1)} = \frac{27{,}380.16 - 5661.4}{76 - 20} = 590.0279\text{, so } s = 24.2905.$$

For variation: when $n = 3$,

$$UCL = 24.2905 + \frac{3(24.2905)\sqrt{1 - (.886)^2}}{.886} = 24.29 + 38.14 = 62.43 \text{ ; when } n = 4,$$

$$UCL = 24.2905 + \frac{3(24.2905)\sqrt{1 - (.921)^2}}{.921} = 24.29 + 30.82 = 55.11.$$

For location: when $n = 3$, $430.65 \pm 47.49 = 383.16, 478.14$, and when $n = 4$,
$430.65 \pm 39.56 = 391.09, 470.21$.

Math 205
C Morgan
3 hours